FACING CATASTROPHE

FACING CATASTROPHE

Environmental Action for a Post-Katrina World

ROBERT R. M. VERCHICK

HARVARD UNIVERSITY PRESS
Cambridge, Massachusetts
London, England
2010

Library of Congress Cataloging-in-Publication Data

Verchick, Robert R. M.

Facing catastrophe : environmental action for a post-Katrina world /
Robert R. M. Verchick.

p. ; cm.

Includes bibliographical references and index.

ISBN 978-0-674-04791-4 (alk. paper)

1. Environmental policy—United States.   2. Environmental protection.
3. Emergency management.   4. Disasters—Risk assessment.   I. Title.

[DNLM: 1. Disasters—prevention & control.   2. Environmental
Monitoring—methods.   3. Disaster Planning—methods.
4. Environmental Health.   5. Public Policy.   WA 670 V481f 2010]

GE180.V47 2010

363.340973—dc22   2010003367

For my sons:
Reed, Ty, and Luke

CONTENTS

ACKNOWLEDGMENTS

I am grateful to many people for their help in writing this book.

Elizabeth Knoll of Harvard University Press wisely encouraged me to broaden the scope of the book and helped me envision the contours of a quickly developing field of law. Her challenging questions and sharp eyes improved the book in countless ways.

Dean Brian Bromberger generously provided his encouragement and support throughout the project. I received critical financial support from the Loyola University New Orleans College of Law and from the Louisiana Board of Regents Endowed Chairs for Eminent Scholars Program.

I have given presentations about many topics in this book and have received many helpful comments. These events include programs organized by the University of Arizona, the University of Colorado, the University of Copenhagen, the University of Florida, Georgetown University, the University of Georgia, the University of Houston, the University of Oregon, Stanford University, and Tel Aviv University. I also benefited from discussions in several classes I have taught on disaster law, including a course jointly sponsored by Loyola University New Orleans and Georgetown University.

I prevailed upon friends and colleagues to critique drafts of the manuscript, in whole and in part. Their insights greatly improved the book. For their efforts, I thank David Adelman, Dan Farber, Eileen Gauna, Rob Glicksman, Steve Higginson, Alexandra Klass, Christine Klein, Doug Kysar, Nancy Levit, Catherine O'Neill, Sid Shapiro, and Karen Sokol.

Others endured repeated conversations about my research, teaching me more about relevant topics or helping me to frame and sharpen my ideas. For that I thank Vicki Arroyo, Jim Chen, Erik Christensen, Lee Clarke, Alyson Flournoy, Lisa Heinzerling, Andrea Davis Henderson, Johanna Kalb, John Lovett, Jim Salzman, Mark Schleifstein, and Wilma Subra. Doug Kysar deserves special thanks for sharing with me his research on the

"Standard Project Hurricane" exercise and for referring me to "The Lakes of Pontchartrain."

I am thankful for an outstanding team of research assistants: Laura Ashley, Tiffany Crane, Whitford Remer, Sharon Stanley, and Samuel Steinmetz. Nona Beisenhurz, one of Loyola's top law librarians, provided exceptional services, identifying and locating sources in myriad academic fields. Rhonda Townson, Mearidith Darbonne, and Lindsey Reed provided excellent secretarial support from start to finish. Thanks also to the patient staff at Café Luna (Ernesto Montano, proprietor), where much of this book was written.

Of course, this book would have been impossible without the support and understanding of a loving family. For their perpetual encouragement, I thank my wife, Heidi Molbak, and my three sons, Reed, Ty, and Luke, of whom I am so proud.

Finally, I would like to thank the people of New Orleans, whose bright spirit and hard work have inspired every page of this book.

FACING CATASTROPHE

INTRODUCTION

O n a clear November morning in 1755, the city of Lisbon was rocked to its core by a titanic earthquake. The event, now believed to have approached magnitude 9 on the Richter scale, was followed by a tsunami and multiple fires. These forces all but demolished that fabled city, along with many settlements nearby.[1] The number of city residents believed to have been killed is 30–40,000; another 10,000 may have lost their lives in Spain and Morocco. Lisbon also lost 85 percent of its buildings, including famous libraries and ornate palaces. Oliver Wendell Holmes, Sr., father of the famous American jurist, captured the scene decades later:

> The ruins of Lisbon burned for six days. . . . The city, according to one observer, was reduced to "hills and mountains of rubbish still smoking." A Mr. Braddock, seeking higher ground after experiencing a sea wave that accompanied the noon shock, described victims with "their backs or thighs broken, others vast stones on their breasts, some lay in the rubbish and crying out in vain . . . for succour." Streams of refugees were fleeing the city, and the earth was not yet quiet.[2]

The Lisbon Earthquake was not the worst disaster to befall a European city, but it was a game changer. Often called "the first modern disaster," the event roused many from a complacent theodicy that had allowed them to grow too comfortable with the aristocracy and vague notions of fate. During the response and recovery phase, citizens demanded more of government and began seeing themselves as agents of change in their environment. In response, the nation's prime minister immersed himself in the practical details of reconstruction and launched one of the first scientific inquiries into the mechanics of earthquakes. Zoning rules were imposed, as were Europe's first building codes for seismic events. The catastrophe also strongly influenced the philosophy of Enlightenment thinkers. Voltaire used the earthquake in *Candide* to attack the "all's for the best" mentality peddled by Gottfried Leibniz. Immanuel Kant's meditations on the "sublime" (an investigation of boundless awe), which would one day influence the preservationist movement in the United States, were in fact inspired by his earlier study of the Lisbon Earthquake.

In April 1927, the great Mississippi River overflowed its banks from Illinois to the Gulf of Mexico. For six weeks more than 20,000 square miles lay submerged beneath as much as thirty feet of water.[3] An estimated 500,000 to 1,000,000 people lost their homes, and nearly 1,000 lost their lives. The consequences led secretary of commerce Herbert Hoover to take on a massive flood relief effort, establishing a national role for the region's redevelopment and setting the stage for his election to the U.S. presidency. A year after the disaster, the federal government took charge of all flood control projects on the lower Mississippi, and in 1936 it took over projects on *all* federal waterways. The devastation and poor treatment of blacks in the aftermath of that flood contributed to the demise of southern sharecropping, the migration of African Americans to the industrialized North, and an invigoration and dissemination of the famous Delta Blues.

On August 29, 2005, Hurricane Katrina, the costliest disaster in American history, changed the game again. The storm's surge plowed across the Gulf Coast from central Florida to Texas.[4] New Orleans was the hardest hit. The surge broke through the federal levee system in more than fifty places, swamping nearly 80 percent of the city. At least 1,836 people lost their lives in the hurricane and subsequent flood (828 in New Orleans), making it one of the deadliest catastrophes in American history. Experts estimate the storm was responsible for more than $81.2 billion in damage.

Katrina is often called a "natural" disaster, but the damage in New Orleans was mainly the result of a massive engineering failure. By all expert accounts, Katrina's surge was *within* the federal levee system's intended design specifications. Two professional independent reviews and one by the U.S. Army Corps of Engineers itself have attributed the flooding of New Orleans to mistakes in levee design and construction, stretching back over decades and under the supervision and control of the federal government.[5] The storm and its aftermath caused Americans to question many aspects of hazard protection as well as of government in general—from environmental protection to race relations, from engineering design to the responsibilities of national government. Many observers outside our borders wondered what the Katrina disaster might say about the importance of coastal ecosystems, the effects of sea-level rise, and the social effects of environmental catastrophe.

Three Environmental Lessons

This book hopes to make sense of many of these issues for the purpose of preparing the United States and the international community for the next "Big One," wherever it may be. Having lived through Katrina as a resident of New Orleans and studied its consequences, I believe our disaster laws and policies are in need of substantial reform. As an expert in environmental policy and law, I find that many of the most important lessons we can learn about preparing for and responding to disaster come from that discipline. That is hardly surprising, given that environmental law has long focused on imagining the worst that might happen to a community or landscape and then crafting efficient and fair ways to avoid it. What is more, so much of disaster mitigation involves these very aspects of environmental protection, from the stewardship of natural storm barriers like wetlands to protecting people's health and welfare in the neighborhoods where they live.

Based on Katrina and other similar catastrophes, we might organize the lessons of environmental protection into three simple commands that can guide policy-makers in preparing for future catastrophes, whether in the United States or the rest of the world.

The commands are: Go Green; Be Fair; and Keep Safe. "Going green" means minimizing physical exposure to geographic hazards by preserving natural buffers against them and integrating those buffers into artificial systems like levees or seawalls. Coastal wetlands, for instance, dampen storm surge and can increase the effectiveness of levees and seawalls. Healthy first-growth forests help contain the spread of wildfires and protect against floods and landslides. Going green also means respecting the limits of natural geography by discouraging new development in areas that expose people and assets to unreasonable risk.

"Being fair" means looking after the public health, safety, and environment in ways that promote distributional fairness and that do not increase personal and social vulnerabilities. In nearly every disaster, it is the poor and other vulnerable groups who suffer most. Disaster response policies must pay particular attention to the needs of the poor, racial minorities, women, children, the elderly, and the infirm. More generally, reforms in health care, housing, employment, and education are needed to reduce the preexisting vulnerabilities of these struggling groups.

"Keeping safe" means assessing and managing risk in ways that capture the full spectrum of values at stake in the context of preparing for catastrophic harm. Too often, important environmental or engineering decisions are made according to economic models that emphasize cost savings or commercial development but downplay public safety. In designing the New Orleans levee system, for instance, the Army Corps regulations allowed engineers to consider the "benefit" to real estate developers of opening new land for development but not estimates of avoided fatalities. Had the latter been considered, a higher standard of safety (and higher government expenditures) could have been justified. The levees that protect thousands of people in the Sacramento-San Joaquin River Delta in California share the same design limitations. Stronger precautionary ideas should be applied to many areas of disaster preparation, including storm modeling, municipal building codes, and land use planning.

Disaster Is Everywhere

It would be one thing if natural hazards threatened only a handful of places. But that's not the case. Disasters come in many varieties and strike in many places, from earthquakes in northern California to wildfires in Montana, from massive floods in the Great Plains to hurricanes on the mid-Atlantic coast. In the spring of 2008, a swarm of tornadoes and storms hit the Midwest and the East Coast, demolishing homes, flooding communities, and sending hundreds of people heading for the hills.

Every year in the United States, events like these cause hundreds of deaths and cost billions of dollars in disaster aid, disruption of commerce, and destruction of homes and critical infrastructure. Although the number of lives lost to natural hazards has generally declined, the economic cost of major disaster response and recovery continues to rise. Every ten years, property damage from natural hazards in the United States doubles or triples. Only Japan suffers more economic damage from natural hazards.

There are many reasons for this trend. Perhaps the most obvious is that the population is growing and expanding into areas that are more prone to natural hazards, like open coasts, forested areas, or scenic hillsides.[6] Global warming also plays a role, resulting in drier, more fire-prone forests, higher sea levels, and according to many experts, more powerful storms. As these trends continue, such events will only get worse.

One of this book's central arguments is that the values of natural infrastructure, fairness, and safety have eroded over the years under a tide of market-based, neoliberal attitudes that have weakened the role of government and shielded it from accountability. Government must assume a stronger regulatory role where natural infrastructure, distributional fairness, and managing risk are concerned. The federal statutes that govern environmental impact assessments, wetlands development, air emissions, use of toxins, and flood control, among other matters, must be reshaped (through regulation or legislation) in acknowledgment of this fact. Governmental decision-making must become more participatory and transparent. The possibility of environmental catastrophes teaches that precaution must become a more prominent—and at the same time workable—feature of many environmental regulations. This book shows how these lessons might be implemented across the board in national and international contexts.

This Book's Approach

Environmental law is an applied discipline. What is valuable is what works. Knowing what works requires some knowledge of the social and natural systems at play. So while my focus is on law and public policy, this book will frequently draw insight from the social and natural sciences. Part I, for instance, incorporates the latest scholarship from ecology, economics, and property theory. Part II includes aspects of sociology and cognitive psychology. Part III borrows ideas from behavioral economics and game theory. Few people can be expert in all of these fields. The key is to find intersections and to create an agenda to which people from many fields can contribute. This book aims to help shape that agenda.

The documentary photographer Henri Cartier-Bresson believed that the more specific a thing is, the more universal it becomes.[7] The maxim holds true for policy development as well as for photography. Because the best way to imagine effective reforms in disaster law is to study specific examples, I will refer to several well-known catastrophes that have occurred in the United States and elsewhere, including San Francisco's Loma Prieta earthquake (1989), the Asian Tsunami (2004), and Cyclone Nargis in Myanmar (2008). But perhaps no catastrophe has more to teach American policymakers than Hurricane Katrina. That is partly because of its scale, as well as the considerable amount of data now available for study. But it is also because, as a relatively recent event, Katrina's secrets have not all been unlocked. Thus I will sometimes put parts of the Katrina story under the microscope in search of universal lessons. This is not really a book *about* Katrina, but Katrina is central to its analysis.

Finally, a word on definitions. Most people think of disasters as events characterized by outsized speed, scale, and surprise. But after that the definition gets blurry. The Red Cross and Red Crescent follow the U.N. International Strategy for Disaster Reduction, which defines *disaster* as a "serious disruption of the functioning of society, which poses a significant, widespread threat to human life, health, property or the environment, whether arising from accident, nature or human activity, whether developing suddenly or as a result of long-term processes, but excluding armed conflict."[8] The Stafford Act, which governs federal assistance in the United States in times of "major disaster," defines that term as "any natural catastrophe (including any hurricane, tornado, storm, high water, winddriven water, tidal wave, tsunami, earthquake, volcanic eruption, landslide, mudslide, snowstorm, or drought), or, regardless of cause, any fire, flood, or explosion . . . which . . . causes damage of sufficient severity and magnitude to warrant major disaster assistance."[9] Law professor Dan Farber prefers to define disasters in terms of the kind of legal and governmental responses they demand. He speaks of disasters as requiring a suite of risk management techniques that includes mitigation, emergency response, compensation, and rebuilding.[10]

Because of my focus on environmental policy, I have narrowed my discussion to a subset of the field. I am concerned with what people generally think of as "natural" disasters like hurricanes, floods, earthquakes, fires, and other similar events. That term is a misnomer, to be sure. While such events include an obvious natural component, the damage they cause almost always involves human action, too, whether an engineering flaw, imprudent land-use planning, or a chaotic emergency response. Anthropogenic climate change has the potential to influence a wide range of disasters, as already mentioned. This is not a book about climate change, but nearly every issue I address is backlit by its presence.

For reasons of space, I also confine the book to disasters that occur suddenly and have limited duration. This leaves out famine and drought, as well as other devastations. While my three simple commands may apply to incremental, long-term events, my analysis will not directly address them. And for those interested in asteroid collisions or the earth-crushing aspects of particle physics, I give these things only the briefest mention, and not until Part III. Finally, I use the terms *disaster* and *catastrophe* somewhat interchangeably. When I mean something more specific, I will indicate as much.

The Second Line

A well-known tradition still seen on the streets of New Orleans is the jazz funeral. A special version of that tradition was, in fact, staged during Carnival in 2005 to commemorate the many lives lost to Hurricane Katrina. The jazz funeral has two stages. First there is a melancholy, dirge-like march, from the church to the cemetery. After the internment comes the "second line," which opens with a scream of trumpets followed by a parade of high-stepping dancers, syncopated percussion, and a rainbow of fringed umbrellas and swooping feathers. The idea is timeless: death is loss and death is sad; but out of death comes *hope* (that "thing with feathers," as Emily Dickinson wrote) and the promise of a brassy resurrection. This book celebrates the beginning of that second line.

I

GO GREEN

1

NATURAL INFRASTRUCTURE

Most infrastructure that we talk about is the gray kind—dams, turn-pikes, and the like. But some infrastructure is green. Indeed *most* of it is. As I will show, green infrastructure, like swamps, forests, and icy sea-walls, does more to protect us from disaster than most people realize. Preserving these features should be a no-brainer, for nature is often stronger and cheaper than many of the things we build to protect people from floods, fire, and landslides. But the law generally does not protect people or their property by preserving the natural infrastructure. Why not? And how could law be reformed to do that job?

Chapter 1 sets the stage for this inquiry by describing the many services nature provides and arguing that they should be seen as a kind of public infrastructure. Seeing nature as infrastructure encourages us to see more value in it. This helps justify a larger role for government in monitoring and maintaining it. For illustration, this chapter focuses on the many services provided by Louisiana's coastal wetlands, which can teach us lessons about ecosystem protection. Chapter 2 builds on these insights and catalogs the many ways that ecosystems throughout the country and all over the world protect human communities from floods, earthquakes, fires, and other ca-tastrophes. Understanding this reliance not only helps us understand the importance of green infrastructure for disaster mitigation but suggests ways we might reimagine government's role in maintaining them. Chapter 3 moves from economics and ecological science to economics and political science, attempting to identify the conceptual reasons that our laws do not protect natural buffering systems as well as they should. Chapter 4 builds on this examination to illustrate some of the ways that environmental laws and property regimes fail to attach proper value to the relevant natural features and recommends specific reforms in public stewardship, regulation, eco-nomic incentives, and the private law of property that would enhance the natural systems we depend on in times of crisis.

"The Marrow of Nature"

First-time visitors to New Orleans acquaint themselves with the city in various ways. They sup turtle soup in a five-star restaurant, groove to electric blues in the Marigny, or swallow oysters at Casamento's. If your mind's on history, you will drop by William Faulkner's onetime home (now a bookstore) or wander through the Sala Capitular, on Jackson Square, where the Louisiana Purchase was signed. Nowadays a necessary but painful tour of the "Lower Nine" might also land on the itinerary. But visitors almost never do what I recommend as the first step toward understanding the "City That Care Forgot": hop in a rented canoe and paddle through the swamps.

Lean forward and imagine pulling a boat through Louisiana's Barataria Preserve, which is part of the Jean Lafitte National Historic Park and Preserve, less than an hour's drive from New Orleans. Imagine paddling there in the early morning, down a waterway called Bayou des Familles. There you are surrounded by more than 20,000 acres of bayous, forests, marsh, and swamp.[1] Some 2,000 years ago, this course was a major distributary of the Mississippi River, depositing millions of tons of sediments that now make up this part of the Delta. The bank from which you launched your canoe is some of the newest land in North America. But ironically, the setting looks ancient, even prehistoric—as if the head of a brachiosaurus is looming somewhere within the forest's lush canopy.

Enormous live oaks and other hardwoods crowd each other for space on a ridge of land just past the bank. Dripping with Spanish moss, their python-like limbs shade an understory of fern and palmetto, along with a menagerie of things that slide and crawl. This ridge, called a "natural levee," descends gradually into swamp, where water stands for most of the year. Here forests of tupelo gum and bald cypress take over, the root-like "knees" of the latter poking out of the water like dragon's teeth. The water itself is not what most people are used to. It's not blue, it's green—or more like brown-green. And it is covered with small bits of vegetation, from small leaves to duckweed, all of which floats along at an almost imperceptible rate. Navigating through this mixture is not difficult but is harder than you might think, like paddling through a tureen of lentil soup.

In addition to vegetation, you'll notice wildlife, although perhaps not immediately. People usually see the birds first. Like John Audubon, you'll marvel at the colorful array of green herons, snowy egrets, and other waterfowl. Look closely and you might notice the mud shelters of crawfish along the bank or the bubbles of a snapping turtle pulling herself along the water bottom. Tree frogs are a common sight, as well as a surprising variety of lizards and snakes. In my visits to the park during both day and night, I've also seen rabbits, squirrels, striped armadillos, butterflies, big-eared bats, and (on one spring night) an obviously confused house cat. Thanks to the Endangered Species Act, the American alligator is also a familiar sight, whether sunning itself on shore or floating like a log in the green-brown muck. If you paddle far enough (and assuming no obstructions) you will eventually slip out of this confined landscape and into an apparently endless stretch of flat, treeless marsh. Here, with perhaps a flock of terns overhead, is the eastern edge of one of the largest expanses of freshwater marsh in the world.

All of this looks like wilderness, but it's not. For centuries the Barataria region hosted a considerable amount of human ambition, manifested in a wide variety of forms, including prehistoric settlements, colonial farming, plantation agriculture, logging, commercial trapping, fishing, and oil and gas exploration. Back in the swamp, on an adjoining waterway called Bayou Coquille, you might have discerned a patch of artificial mounds rising gently from the banks. Those are middens, the ancient refuse heaps of American Indians who settled here over 2,000 years ago. These mounds, which are packed with sea shells (*coquilles* in French), offer some of the only evidence of the region's ancient settlements, many of which thrived on clams and other shellfish.

Beginning in the early 1700s, people from Europe, Africa, and Asia later built permanent settlements in the region.[2] Jean Lafitte, the pirate and privateer for whom the park is named, is thought to have evaded enemies by hiding in these forests. Through a veil of vines and Spanish moss, you might discern evidence of house foundations, "fish camps," furrows, ditches, and even plantation fields. In the 1770s, families of Canary Islanders, known as Isleños, settled briefly near Bayou des Families (meaning "of the families") until floods and hurricanes drove them away. But others would take their place, clearing the higher land for sugarcane and stripping the swamps of bald cypress, whose beauty was surpassed only by its strength as a building material. The current cypress forest, which is second or third growth, is spindly compared to what once existed. (The ancient stump of a first-growth cypress—eight feet across—will prove the point.) If you find yourself on a waterway that is unusually straight or deep, you may have happened onto an old logging "cut," once used by logging companies to float cypress out of the swamp. Or you could be on a canal once used for oil and gas exploration, an important industry that dates back to the 1930s. The occasional "No Hunting" sign will remind you of the swamp's other uses, as will another sign familiar to paddlers in these parts: "Do Not Anchor or Dredge, Natural Gas Pipeline." Oil and gas have been big-ticket items in modern times, and perhaps no more so than today. But real estate is big, too, as suburbanites move farther away from the city in search of cheaper land and more space. They find both, as luck would have it, in new tract-mansion developments just outside the Barataria Preserve where swaths of longleaf pine once stood.

Paddling through the bayous, your eyes naturally focus on *things*. Using a little thought and imagination, you might also consider the *processes* that keep this region going. We've already seen the slight changes in land elevation and noted the role of sediment-filled rivers in shaping these features. The sediment, rich in nutrients, forms habitat for hardwoods and other vegetation, which, in turn, provides habitat for birds, reptiles, and rodents. In the delta, water and soil determine everything—where things grow, what survives, and what dies. If the water is too fresh, some shellfish die; if it's too salty, Gulf predators move in. Oaks and maples thrive on the delta's higher and drier land, while cypress prefer transitional waters. Look back at the wake of your boat and see how those expanding ripples are gradually absorbed by the tangle of roots, sand, and twigs that make up the shore. That barrier protects these fragile banks from erosion. Notice the distant row of cypress trees, all snapped at about the same height, like broken telephone poles. Imagine the impressive force they absorbed as the winds of Katrina and Rita charged through this landscape.

In a swamp, it's hard to ignore the presence of light and heat. Through the process of photosynthesis, sunlight enables plants to transform water and carbon dioxide into essential nutrients. The intense sunlight feeds the forest's thick canopy. Yet the sun's heat would cook the lizards and frogs if they could not lower their body temperatures by slipping into the same canopy's cool shadows. Smaller plants, like those spiky palmettos, also benefit from the canopy's shade. When sunlight hits the forest floor, faster growing exotic plants can take over, eating up the levee's desirable real estate and transforming the habitat. Indeed, this is exactly what is happening in the Barataria Preserve, roughly 60 percent of whose forested canopy was torn apart by the 2005 hurricanes.

But enough of Mother Nature. Many of these environmental processes affect us too. By creating habitat for plants and animals, the swamp's life-support functions nurture things we can use or consume, from the millions of clams that supported ancient coastal people to the cypress logs used to build much of New Orleans. And don't forget the raw oysters and turtle soup, which as visitors learn even before leaving the airport, trace their beginnings to the state's famous swamp. Then there is the whole process of water control, those ridges and hammocks that buffer surge and soak up overflow. Remember also the cypress and oaks that dampen high winds pounding onto the shore.

Personally, though, I would not keep coming back to these waters—with students, friends, or family in tow—were it not for another service this landscape continually provides: an undiluted sense of awe and wonder. It's simply impossible to paddle through these soupy waters without losing yourself in the mystery, the strangeness, and the wild-ness of it all. You can call it a "spiritual journey," or "recharging the batteries," but encounters with ecosystems as wondrous as these fulfill basic needs important to personal reflection, socialization, and human culture.

It is sometimes forgotten that Henry David Thoreau, the philosopher-ecologist who wrote so movingly of sparrows, "sprout-lands," and ice, seemed most at home in a swamp. "The very sight of [a] half-stagnant pond-hole, drying up and leaving bare mud," he once mused, "is agreeable and encouraging to behold, as if it contained the seeds of life, the liquor rather, boiled down."[3] But it was even better to be *inside* a swamp, enveloped so completely by water and vegetation that notions of geography and scale dissolved. "I seek the darkest wood, the thickest and most interminable and, to the citizen, most dismal swamp," he wrote in his essay "Walking." "I enter a swamp as a sacred place, a *sanctum sanctorum*. There is the strength, the marrow of Nature."[4] Imagine having the "luxury," as he put it, "to stand up to one's chin in some retired swamp a whole summer, scenting the wild honeysuckle and bilberry blows, and lulled by the minstrelsy of gnats and mosquitoes!"[5]

Mosquitoes excepted, modern scientific research confirms the human need for direct encounters in the natural world. In *Last Child in the Woods*, journalist Richard Louv argues that many disturbing trends in our children's health, from obesity to attention disorders to depression, can be linked to an absence of nature in our lives.[6] Outings in wetlands like these may actually contribute to the physical and emotional development of our children. Perhaps no one has done more to promote the beauty of Louisiana's wetlands than wildlife photographer C. C. Lockwood, whose books and articles in *National Geographic* and elsewhere have captured the many wonders and ironies of this essential landscape.[7] In 2004, he and his wife, Sue Lockwood, lived for a year aboard a houseboat exploring the Atchafalaya's coastal wetlands and documenting what they found. C. C. kept journals, tracked alligators, and shot thousands of photographs. Sue, a teacher, kept a blog for school children and developed hands-on lesson plans in geography and science for Louisiana classrooms.[8] In a book dedicated to this experience, C. C. writes: "[w]e are all tangled up in a maze of pipelines, industry, and marine commerce that interacts with community, culture, recreation, and tourism. All the while, this hodge podge of human activities depends upon the natural elements, the minerals, the flora, and fauna that make up the bountiful coast."[9] It is, perhaps, a simple idea—the importance of a bountiful coast to human hodgepodge. But as you float beneath Louisiana's tea-stained skies for one last time, consider that things are about to get much more complicated.

A Working Coast

Let's leave the bayou and examine the Louisiana coast from a different perspective. Pull out a map of the United States and with your finger trace the course of the Mississippi River, tumbling from a small glacier lake in Minnesota, flowing through the Heartland, and eventually emptying into the Gulf of Mexico. The last hundred miles are, of course, in southeast Louisiana, an irregular finger of land jutting into the sea. In both form and function, Louisiana's southwestern tip is our nation's intestinal colon. That birdfoot delta at the end drains three-quarters of all the land in the United States. Over thousands of years, the Mississippi has changed its course several times, spraying water and mud across the Delta region like the nozzle of a loose garden hose. Some 2,500 years ago, the river began building a delta in the area that is now New Orleans. Back then the delta had three primary channels, one of which was Bayou des Familles, the subject of our paddling tour.

Since the time of the first native settlements, the Louisiana coastal plain has always been what locals call "a working coast." In the early twentieth century, before oil and before federal levee projects, commercial activity was dominated by the businesses of transportation, fish, and fur. Even then, the growing commercial use of the coast raised concerns among some scientists that the wetland's "all-you-can-eat buffet" had a limit. One of those scientists, the Louisiana naturalist Percy Viosca, Jr., is viewed today as a kind of patron saint of bayou ecology. In the 1920s Viosca became concerned with the toll that all the hunting, trapping, and fishing was taking on the coastal plain. "During the past fifty years," he would note in 1928, "great changes have taken place in this State due to man-made factors which are changing the conditions of existence for our wild life and fisheries from its very foundations."[10] Hoping to support the cause of conservation with hard numbers, he set about trying to quantify the importance of Louisiana's wetlands to the human economy. Focusing mostly on valuing the marketable goods taken each year from the coast, he estimated that the annual value of freshwater fish produced in the wetlands was more than $2.5 million. "Frogs, turtles, freshwater shrimp, crayfish and fresh water mussels" were worth another $1 million by his calculation. Saltwater fisheries (including crabs and diamondback terrapin) produced an annual value of $2 million, in addition to the even larger market for shrimp ($3.5 million). As for the fur industry, Viosca found those skins and pelts generated a whopping $6 million every year.

Using these and other numbers, Viosca estimated that the state's wetlands each year produced goods and services worth more than $20 million (about $250 million today).[11] He was well aware that Louisiana's working coast provided more human benefit than he was able to capture in these crude calculations. He understood, for instance, that the swamps offered opportunities for recreation and quiet reflection, while also inspiring a tight-knit and buoyant culture. More significant, from my perspective, he recognized the central role wetlands play in combating floods and storms. He opined that the thing we could least afford to lose was the "flood protection in the valley lands of our rivers by means of [natural] levees which render the front lands and considerable areas of the back lands reasonably free from annual floods."[12] Thus, Viosca became an early champion of wetlands conservation. Unchecked human activities were, as he put it, "killing the goose that laid the golden egg."[13] To avoid future collapse, he urged that coastal degradation "be considered a state and national problem equal in significance to agricultural development."[14] Viosca's plea to view ecology as infrastructure went in large part unheeded, leaving it for a future generation to consider the damage and plot a reversal.

With this goal in mind, let's flash forward to the present and consider the map more carefully. Louisiana's coastal plain contains one of the largest expanses of coastal wetlands in the contiguous United States.[15] In fact, one-quarter of all the coastal wetlands in the Lower 48 are in Louisiana. The state's wetlands are so big, as my friend Oliver Houck at Tulane Law School likes to say, "You could drop the Florida Everglades into our swamps and never find them again." As we've seen, Louisiana's coastal plain hosts an extraordinary diversity of coastal habitats, ranging from natural levees and beach ridges to large swaths of forested swamps to freshwater, intermediate, brackish, and saline marshes. These features, which nourish wildlife, filter water, and dampen storm surges, help make the coastal plain, to quote the Army Corps of Engineers, one of "the most productive and important natural assets" in the country.

To visualize how coastal marshes provide one of their most important services, storm protection, imagine blasting water through a garden hose at full force onto a cement driveway. The water splashes and surges, fanning out in many directions. Now imagine spraying water from the same hose onto a thick, dense lawn. The difference between the cement and the lawn is the difference between a storm path composed of open water and denuded coast and one composed of lush forests and marsh. Louisiana's coastal wetlands act as vast sponges, absorbing billions of gallons of rainfall and shielding people and property from storms. The effect is impressive, even for city dwellers who have never seen a marsh: by some estimates, every two miles of wetlands south of New Orleans reduces tropical storm surges there by half a foot.[16] For this reason, local environmentalists sometimes refer to the coast as a "horizontal levee." These coastal wetlands are also helping to shield an internationally significant commercial-industrial complex. The area's complex of deep-draft ports, including the Port of South Louisiana, handles more tonnage than any other port in the nation.[17] According to recent figures, "Louisiana led the Nation with production of 592 million barrels of oil and condensate (including the outer continental shelf), valued at $17 billion, and was second in the Nation in natural gas production with $1.3 billion (excluding the outer continental shelf)."[18] In addition, more than 29 percent of the country's crude oil supply and nearly 34 percent of its natural gas supply moves through Louisiana, which, incidentally, also hosts about half of the nation's refining capacity.[19]

In addition to storm protection services, the Louisiana coastal plain provides many other benefits. It offers habitat for countless species, including commercially significant sea life and waterfowl.[20] Fisheries in the Gulf of Mexico provide about 20 percent of all seafood consumed in the United States.[21] Nearly all of that catch is dependent, in some way, on the universe of microscopic plant and animal life first nurtured in the coastal plain. With more than five million birds wintering in Louisiana, the Louisiana coastal plain provides crucial rest stops to migrating birds.[22] Finally, Louisiana's coastal marshes provide services that are vital to water quality. The marshes function as giant "water treatment plants," filtering out vast quantities of nitrogen, phosphorous, and other pollutants from incoming water bodies.[23] Taken together, the many services of Louisiana's coastal wetlands make them a treasure every bit as unique and breathtaking as the city of New Orleans itself. The coast's storm protection, habitat, and water treatment services, while impossible to precisely quantify, surely amount to billions of dollars of commercial benefit per year.[24]

Unbelievably, this titan among wetlands, this biotic and commercial treasure, is disappearing before our very eyes. Since Percy Viosca's enlightened plea for wetlands protection, Louisiana has lost more than 1.2 *million acres* of its "trembling prairie."[25] Today, the Army Corps of Engineers believes Louisiana is losing about 6,600 acres of coastal wetlands per year.[26] Why is this happening? The effect is partly due to natural subsidence. The soft soils of the coastal plain naturally shift and sink over time.[27] But this phenomenon, at best, explains only a small fraction of the loss.[28] The real culprits are human-made: Louisiana's vast network of levees, navigational channels, and oil-and-gas infrastructure. The levee system accelerates coastal land loss by reducing the natural flow of the river's freshwater and sediment to wetland areas, where the lost land would then naturally be replenished.[29] Instead, that valuable water and sediment is funneled down the Mississippi and shot into the Gulf, toward the outer continental shelf, where the formation of barrier islands is impossible.

Louisiana's coastal plain is crisscrossed with a vast matrix of navigational canals, including 10 major navigational channels[30] and literally thousands of smaller access canals serving navigation, allowing oil-rig access, and cradling oil and gas pipelines.[31] Recall that up close, those swamps seemed lusty and vital. But from a helicopter or plane, the view is more depressing — like green shag carpet being eaten by a Pac Man. These channels mean death to nearby marshlands. They cut the land off from the natural flow of water and starve them of nutrients. Often the banks of these channels (which are seldom maintained) crumble and fall away, expanding waterways by many yards over time.[32] The major navigational channels pose their own special threat to flood control by sometimes acting as "hurricane highways," allowing storms to sweep inland past marshland like liquid bulldozers.

In the 1980s, local groups made up of environmentalists, shrimpers, scientists, and businesspeople began pushing for plans to save the coast. One result was the federal Coastal Wetlands Planning, Protection and Restoration Act of 1990. The legislation created a federal and state task force to implement wetlands restoration projects with annual funds of around $40 million. In 1998, state and federal agencies, with the participation of a diverse group of local churches, scientists, environmentalists, and fishers, developed a plan they called Coast 2050, which offered a host of ecosystem restoration strategies.[33] The underlying principles of this plan were to restore or mimic the natural processes that built and maintained coastal Louisiana. The complete plan, to be implemented over the next 50 years carried a price tag of $14 billion, more than twice as much as the Everglades restoration project (nearly $8 billion). Though expensive, Coast 2050 actually seemed a bargain, considering that the cost of doing nothing threatened to exceed $100 billion in lost jobs, lost infrastructure, lost fishing, and increased hurricane damage.[34] But Coast 2050 was never funded. In 2004, hamstrung by climbing deficits, the Bush administration suggested that the Corps lower its sights and focus on only a few projects totaling between $1 and 2 billion. In the end, the Corps did not even receive that amount. The president's 2005 energy bill provided only $540 million for Louisiana's coastal restoration over four years.[35]

Hurricanes Katrina and Rita changed all this—first by tearing chunks of wetlands out of the Louisiana coast, making coastal erosion an even more serious problem, and second by highlighting for the public the essential role that wetlands play in storm protection. I mentioned that the Army Corps estimates that Louisiana loses 6,600 acres (around 25 square miles) of its coast every year under "normal" circumstances. Overnight, Hurricanes Katrina and Rita destroyed *four times* that amount of wetlands, transforming 217 square miles of productive swamp and marsh into open water.[36] Some of the most heavily damaged marshland was located south and east of Lake Borgne, wetlands that once offered protection to St. Bernard Parish, a white working-class community, and to New Orleans's own Lower Ninth Ward.[37] In 2008 Hurricanes Ike and Gustav again raked across this fragile coast, threatening coastal protection and commerce. Rex Caffey, director of Louisiana State University's Center for Natural Resource Economics and Policy, estimates that the state's oyster industry lost $6.3 million, nearly 21 percent of its 2007 revenue, to those storms. The much larger shrimp and crawfish fisheries suffered even more, losing an estimated $30.5 million and

$19 million respectively. Talking to a local reporter, Dave Cvitanovich, an oyster grower and a relative of the family that owns Drago's restaurant explained: "By far, coastal erosion is our biggest problem now. . . . I'm out on the water every day and, God, it's accelerating. There's no slowdown. It's like watching an ice cube melt on your table."[38]

The good news is that public awareness of the problem is at an all-time high. Even schoolchildren will tell you that we are losing a football field's worth of land every half hour and that those football fields are one of the first shields against bad storms. For the first time, the federal government may be getting serious about restoring the coast, understanding that it will take billions, not millions, of dollars. There is still a very long way to go, with many pitfalls along the way. I will talk more about these challenges in later chapters. But the storms of 2005 sent a clear message: in addition to *goods*, the Louisiana coast provides *services*, most significantly storm protection. We can't afford to kill the goose that lays *that* golden egg.

Nature as Infrastructure

People neglect infrastructure at their peril. Americans were reminded of this nearly two years after Katrina when a span of the Interstate 35W bridge in Minneapolis collapsed during rush hour. Dozens of cars plunged 60 feet into the Mississippi River, slamming against one another as they fell. Over 100 people were injured, and 13 died.[39] Days afterward, some journalists expressed shock that such a thing could happen in the United States. But we have known for years that America's infrastructure is crumbling. A 2005 report by the Federal Highway Commission, for instance, described 75,000 of the nation's bridges, or about one-eighth of the total, as "structurally deficient." In 2005, the American Society of Civil Engineers, in its biannual "report card" on infrastructure, assigned the country an average grade of D. Ironically, bridges earned one of the highest marks by category—a straight C—while grades for dams, roads, airports, and drinking water systems fluttered between D-plus and D-minus. As for levees, the Army Corps's national evaluation of levee systems after the New Orleans flood found 122 levees in 30 states and territories to have an "unacceptable" risk of failing.[40] Nearly a third of these levees are in California.[41] Other threats result from a refusal to expand infrastructure to meet current need. For instance, Adam Cohen of the *New York Times* writes that "Georgia's failure to build enough reservoirs has contributed to a water crisis that could cripple metropolitan Atlanta."[42] The American Society of Civil Engineers estimates that it would cost about $1.6 trillion to bring our infrastructure up to acceptable levels. The price tag for bridges, alone, tops $188 million.[43] With federal dollars lacking and states strapped for cash, many public leaders are now looking to the private sector to help maintain existing infrastructure and to build the next generation of roads and bridges.

We don't normally think about it this way, but nature is infrastructure, too. A typical dictionary definition of *infrastructure* refers to "the substructure or underlying foundation . . . on which the continuance and growth of a community or state depend."[44] The definition conjures images of cloverleaf highways and broad-shouldered dams. But the word includes more than things made of steel and concrete—what civil engineers sometimes call *gray infrastructure*. For instance, the U.S. government uses "infrastructure" to describe a wide range of service networks, including those related to communications, public health, energy, and finance.[45] Nature, too, provides a vast array of foundational structures and services on which "the continuance and growth" of a society depend. I've mentioned the flood control features of coastal wetlands. But think also of the protein produced by wild fisheries, the carbon sequestration of forests, or the medicine chest hidden within the world's biodiversity.[46] From now on, I will refer to ecosystem features that act as a substructure for human flourishing—like swamps, forests, and rivers—as *natural infrastructure*, or *green infrastructure*, to emphasize their similarity to the other more obvious things we depend on such as airports, bridges, and hospitals.[47]

Viewing the natural structures and systems as infrastructure can help policy-making in many ways. First and most important, this perspective highlights the many ongoing services natural systems provide and encourages us to find ways to protect what we most need. On some level, civilizations have always recognized the importance of natural infrastructure. The ancient Romans—celebrated for their roads and aqueducts—placed nearly equal emphasis on regulating rivers, bays, and other natural features that were important to commercial and cultural objectives. Early Hawaiian communities integrated concepts of natural services into their system of land tenure, making sure that each tribal family had access to freshwater, vegetation, and beach land.[48] The U.S. Supreme Court signaled the importance of natural waterways when in the 1824 case *Gibbons v. Ogden* it recognized Congress's constitutional authority to regulate navigable rivers for commercial purposes.[49] There are many other examples.

But it was not really until the twentieth century that researchers began trying to separate and quantify nature's goods and services in some comprehensive way. I've already mentioned the early work of Viosca, who was experimenting with ecology and economics around the same time Louis Armstrong was pioneering scat singing. By the 1990s the room would be in full sway, with natural scientists like Edward Farnworth, Robert Costanza, and Gretchen Daily building "ecological economics," a new hybrid discipline.[50] Their purpose went way beyond counting shrimp and valuing pelts. Today's ecological economists are interested in the whole array of ecological services one finds in an ecosystem, from photosynthesis to water filtration to fish on the hook. They seek to *identify* those goods and services, *quantify* their value in terms understandable to policy-makers, and *protect* the many ecosystems we can't afford to lose.

Second, in addition to emphasizing hidden services, an infrastructure perspective helps remind us that natural goods and services come as part of larger, interconnected systems. We do not refer to a collection of crisscrossing highways as "infrastructures" (plural). We use the singular form to suggest the collective, interdependent nature of the whole. Similarly, the American Society of Civil Engineers uses "aviation infrastructure" to refer to the nation's network of airports; we can even speak of the transportation infrastructure to describe all airports, flight routes, train stations, rail lines, buses, roads, and so on. It is a way of making sure we do not miss the forest for the trees.

Third, the infrastructure perspective highlights the open character of these fundamental services. Traditional infrastructures are generally offered on an open-access basis.[51] This means that all members of a community have the right to use infrastructure resources in a nondiscriminatory manner. The Roman aquifers supplied water for all. The development of the public trust doctrine, which will figure in an upcoming chapter, ensured that the public had access to the infrastructure of bays for fishing and rivers for navigation. The United States follows this doctrine, guaranteeing public access to beaches, navigable waters, and hunting and fishing. Roads and bridges are also traditionally treated as open. This does not mean that infrastructure use is free or unregulated. We pay to watch cable television, post letters, and even cross bridges. Some infrastructure networks, like shipping lanes or flight routes, are closely supervised and tightly regulated. But within these parameters, infrastructure is open to all legitimate users in a nondiscriminatory way, regardless of the individual's identity or end use.[52] Generally speaking, we keep infrastructure open for two reasons: utility and fairness.

Open access increases utility because it encourages broad investment, innovation, and experimentation with common resources, maximizing the possibility of public utility. Open access is a theory of synergy, in which the whole of user productivity is greater than the sum of the infrastructures' inputs. Take the humble highway. It would be possible, of course, to restrict roads only to those drivers most likely to maximize social output; we could assign a government regulator or competitive market to make such choices. But generally speaking, this does not work in the long run, because the cost of choosing "winners" (which includes the cost of getting it wrong) is higher than opening roads to everyone, including the "losers," whose contributions may be real but hidden.

Concern for social utility is one reason people get uncomfortable with private ownership of critical infrastructure. They fear that private owners will manage resources in ways that maximize their individual profits at the expense of the public's spillover benefits. When Louisiana residents learned recently of preliminary discussions to lease management authority over the Lake Pontchartrain Causeway (the longest bridge in the world) to a multinational engineering firm, they rebelled, and civic leaders quickly scotched the idea.[53] Residents were afraid a private owner would raise tolls, increasing its profits and discouraging public use. For the same reason, many people oppose private control of many aspects of the Internet.

The fairness perspective is based on the normative values of liberty and equality. There are some goods or services so fundamental to human liberty that they should be available to some minimal degree to everyone. Or at the very least, they should be available to everyone on an equal basis. To use the causeway example, some feared that private ownership would compromise access to a fast evacuation route. They thought it might be one thing to allow private management of a commercial thoroughfare, but the importance of a major escape route essential to public safety implied to some people a special need for exclusive governmental control. The open access theory will figure prominently in Part II.

Finally, thinking of nature as infrastructure emphasizes the point that many of our natural systems, no less than streetcars or sewage plants, require protection, by means of *monitoring* and often *maintenance*. In the case of gray infrastructure, the system would often be too costly to build again from scratch; in the case of green infrastructure, such rebuilding may be all but impossible.

2

OUR INCREDIBLE SHRINKING
INFRASTRUCTURE

If you think about disasters long enough, everything seems like a sign. Back in October 2005, after Katrina, I was living temporarily in Houston and making long weekend commutes to check on my house in New Orleans. For miles along the way, all you could see were ruined buildings and the skeletal remains of billboards and fast food signs. I was driving through this desolation one day when my eyes were drawn to a red light on the dashboard saying, "Maintenance Required." With teeth clenched and driving no more than thirty-five miles per hour, I resolved to keep going. Eventually I made it to a medium-sized town with a car dealership. A young mechanic took a look at my car and after identifying a bad alternator promised a speedy repair. He was a nice kid, I remember thinking, with an earnest face and efficient hands. He kept a Bible on his desk, too, wedged between a lamp and a stack of auto manuals. When he finished the job, we went to his desk. He found a pen that worked and wrote up the invoice. He knew I was an evacuee, and he said he was sorry about that. He even offered me a "Katrina discount," which I quickly accepted. "Well, good luck, then" he said, after I had thanked him. "Don't worry too much." Then his voice lowered as if confiding a secret, "The Bible says when the world comes to an end, it will be by fire, not flood."

End of the world or not, nature has plenty of ways to make mischief. This chapter reviews the many ways that natural infrastructure protects people all over the world, from storm-buffering mangroves in Florida to fire-suppressing rainforests in Sumatra. That part of the story is inspiring. The troubling side is that many of nature's most important fortifications are in danger of falling apart. There's a red light flashing on the global dashboard, and it says "Maintenance Required."

In 2001 the U.N. Environmental Programme proposed a global interdisciplinary project, the Millennium Ecosystem Assessment, to evaluate the health of the world's natural infrastructure. Costing over $20 million and involving 1,300 scientists, this assessment was called in 2005 "the boldest move to embrace and understand ecosystem services to date."[1] The effort produced more than a dozen volumes of technical reviews and policy papers. But the bottom line, as set forth in a 2005 "synthesis" report, was this: more than 60 percent of surveyed ecosystems "are being degraded or used unsustainably," and without intervention the situation will only get worse.[2] Systems or services on the critical list included fisheries, water supplies, water purification, climate regulation, aesthetic enjoyment, and, significantly, "natural hazard protection."[3] On this last concern, the report explained: "People are increasingly occupying regions and localities that are exposed to extreme events, thereby exacerbating human vulnerability to natural hazards. This trend, along with the decline in the capacity of ecosystems to buffer from extreme events, has led to continuing high loss of life globally and rapidly rising economic losses from natural disasters."[4]

Around the same time, other organizations began studying the role of natural infrastructure in reducing the impacts of naturally triggered disasters. A 2001 report by Worldwatch Institute examined disasters around the world and their relationship to ecosystems services. Ecosystems, the report concluded, serve as "shock absorbers" against disaster. Many of these systems are more effective and cheaper to maintain than the human-built kind.[5] Two years after the 2004 Asian Tsunami, the World Conservation Union published its study *Ecosystems, Livelihoods, and Disasters*, which examined recent disasters in Asia and Latin America and proposed a "comprehensive approach to disaster management" that would include both humanitarian and ecological objectives.[6]

These reports, and others like them, all agree on a few things. To begin with, despite the efforts of thousands of scientists around the world, there is still a lot that we don't know about the relationship between disaster and natural infrastructure. The protective role of natural systems depends heavily on situational factors—the kind of disaster, the geography, the exposed population, and other details—and there is a need for more place-based examination. Thus, as an "essential first step" to framing disaster policy, Janet Abramowitz of Worldwatch recommends "[i]dentifying and delineating natural resources (like watersheds and floodplains), hazards (such as flood zones), vulnerable infrastructure (such as buildings, power lines, and bridges), as well as vulnerable communities and resources—and doing so at scales that are meaningful to communities and decisionmakers."[7] A related problem is that much of the data we do have is scattered throughout academic disciplines and throughout many countries; and there are not enough people who are tying all of this learning together. The biologist counting polyps on the Great Barrier Reef doesn't talk with the fire suppression expert in Yellowstone National Park; and neither of them, I'd wager, chats much with environmental lawyers or policy-makers. In a small way, this chapter tries to pull together some material from various disciplines and countries to show the importance of synthesizing the information we already have so that we can make better laws and polices in the future.

A second insight from these studies is that small efforts can add up. Newspaper reports—and some examples in this chapter—seek reader interest by focusing on dramatic stories in which thousands of lives are saved because of a providently placed forest or coral reef. We can learn from that. But small ecosystems that reduce risk by small amounts can also have dramatic effects over time, particularly where hazards recur regularly. We should not forget the small villages and the ecosystems that serve them.

Finally, these studies make the important point that disaster mitigation is a two-way street. Of course, we should protect ecosystems that shield communities from harm. But we also need to think hard about why those communities are exposed to harm in the first place. Why are they located where they are? Could they have been built differently? What should future planning look like? As Janet Abramowitz suggests, going green does not mean tempting fate. "[W]e continue to put more people and more 'stuff'— buildings, bridges, cities, and power plants—in harm's way and have weakened nature's ability to mitigate hazards. Equally important," she writes, "is understanding that just as our development choices have made the threats worse, we have the power to make better choices."[8]

Natural infrastructure helps us deal with disaster in two ways. First, nature blocks the punch by slowing or redirecting natural forces that are headed toward people. After the impact, nature helps again by providing goods and services important to physical and economic recovery. I'll examine each role in turn, spending more time on the "punch" because of the diverse and comparatively concrete examples available in the research. The latter half of this chapter takes up the issue of recovery.

Blocking the Punch

Nature blocks the force of disaster through an array of ingenious and complex systems. As the World Conservation Union has affirmed, "[w]ell-managed ecosystems can mitigate the impact of most natural hazards, such as land-slides, hurricanes, and cyclones."[9] Such protections exist to some degree for all major categories of naturally triggered disaster, which I group here for convenience as disasters of wind and water; earthquakes and landslides; and fire.

WIND AND WATER

Destruction from wind and water accounts for most naturally triggered di-sasters. That should not be surprising, given that storms are increasing, sea levels are rising, and half of the world's population lives near the coast. Luckily, ecosystems throughout the world offer an array of services that help human societies keep the dangers of wind and water in check. Generally, those services relate to foundational soils, forests and wetlands, barrier is-lands, and coral reefs.

Soil is an obvious, but easily missed, component of protective infrastruc-ture. Soil, after all, is what holds everything up, and if it's high enough, keeps everything dry when floodwaters come. In a land of high water, elevation is destiny. In New Orleans, stronger soils and an elevation 2 feet above sea level assured that the historic French Quarter survived nearly untouched. In other parts of the city, with canal flood walls design flaws failed in part because of overly soft soils at their foundations.[10] Once breached, the walls allowed estuarine waters to sink dozens of neighborhoods, most of them built on land several feet below sea level.

People build on soft lowlands for the same reasons they choose to live in earthquake country or on the sides of volcanoes—along with a big down-side comes a big upside. The same geologic plates that have buckled up beneath the San Andreas and Haywood faults are also responsible for shap-ing the commodious and magnificent San Francisco Bay. The same volca-nic ash that can asphyxiate an Indonesian village also enriches farm soil with essential organic nutrients. For thousand of years, people have settled near deltaic systems despite their fragile character. The Nile delta, the Yangtze delta, and the Bengal delta were all cradles of civilization because such deltaic rivers and streams provide a great wealth of infrastructure-based services, including rich soils, animal habitat, fisheries, cultural and spiritual inspiration, and channels of navigation. There is always flood and pesti-lence, too, but for millenia the bargain has seemed worth it. That might be changing. Today scientists tell us that these and other superdeltas are, for the first time in memory, gradually beginning to sink.

The reason relates to an ecosystem process. Left on their own, river deltas are always expanding and, to a lesser degree, contracting. A river like the Mississippi or the Nile would carry hundreds of millions of tons of upstream sediment into its delta, depositing gravel and sand for miles along the way. These new landmasses would form barrier islands, wetlands, and giant embankments, like the "natural levees" on which the New Orleans French Quarter now stands. Wind and rain would erode some of these features, and less compact soils would inevitably sink. But the next year, alluvial floods would replace the lost soil, building newer and often bigger castles in the sand. Today, however, the balance between sediment and sea level is out of kilter. In many of the world's largest deltas, upstream dams and canals keep sediments from replenishing the soils. To make matters worse, landowners are extracting surface-supporting water and petroleum from beneath coastal land at a record pace, causing it to sink even faster.

Recently, a team of conservation scientists, led by Jason Ericson of the University of New Hampshire at Durham, studied 40 of the world's largest delta systems and concluded that *all* of them are sinking and face threats of encroaching coastal waters. Nearly 70 percent of these systems are sinking mainly because of sediment starvation, and the remainder mainly because of resource extraction or increased sea-level rise. Populations most threatened include not only those on the deltas of the Mississippi, Nile, and Bengal rivers but also the tens of millions living on India's Godavari delta, China's Yangtze delta, and the Mekong delta in Vietnam. What's more, all of these regions, except for the Nile, are prone to violent tropical and subtropical storms.[11]

In the Bengal delta, where storm surges can top 30 feet, the land is sinking an inch per year. Millions of gallons of surface-supporting groundwater are pulled out of the ground each year to be used by the region's nearly 3.5 million people. Surge and subsidence make a bad combination. The Bengal Delta was the site of the two deadliest storms of last century: the first killed about 300,000 people in 1970; the second killed 138,000 in 1991. More recently, in 2007, Cyclone Sidr walloped the delta again, displacing a million people and taking more than 3,000 lives. The Godavari delta, on India's east coast, tells a similar story. The region's 453,000 residents risk losing nearly a quarter of their land by 2050, much of it due to subsidence from groundwater removal. A tropical storm killed more than 1,000 people when it hit the region in 1996. In 1999 a so-called supercyclone inundated a neighboring delta just north of the Godavari, killing some 10,000 people.[12]

Hurricane Katrina showed Americans the danger of such trends. We've already seen the effect that water projects have on sediment loss in the Mississippi delta. Upstream dams cheat the region of land-building silt, while an unforgiving levee system shoots the remainder into the open sea. On top of that, the extraction of offshore oil and natural gas leads to seismic shifts and further subsidence.[13] Scientists and engineers are working hard to develop strategies for rebuilding the lost coasts of Louisiana and other places around the world. (Later I'll tell about the Army Corps's plans to build new islands, spits, and natural levees with millions of tons of imported soil and sand.)

Coastal erosion is not just a concern in steamy lowlands; there are problems up north, too. In 2004, for instance, the General Accounting Office (which, on July 7, 2004, was renamed the Government Accountability Office) reported that 6,600 miles of Alaska's coastline and lowland riverbanks are "subject to severe flooding and erosion."[14] Much of this affects Alaska Native villages, whose roads, homes, and gas stations are slowly washing away. Federal and state officials estimate that 184 out of 213 native villages (or 86 percent) are threatened in this way. Coastal villages, of course, have always endured annual storms and the beach erosion they cause. But climate change is making things worse. In the last decade or so, rising temperatures have delayed the formation of thick barriers of shore ice, which protect coastal communities and help prevent soil loss. This delay now leaves some native villages fully exposed to fall storms. In addition, rising temperatures are also loosening protective layers of permafrost that help hold the land in place. The combination can be devastating. During one storm in October 1997, the barrier island village of Shishmaref—which is less than 1,320 feet wide—lost 125 feet of its beach. Some villages, including Shishmaref, are now planning to relocate, at the cost of several million dollars. The Inupiat village of Kivalina, on Alaska's northwest coast, made news in 2008 when it filed a nuisance claim in federal court against 24 oil, gas, and electric companies to hold them responsible for coastal damage allegedly caused by the defendants' greenhouse gas emissions.[15]

Beyond soil, wetlands and forests also provide important infrastructure for protection against wind and water. Scientists cannot say for sure what amount of damage caused by hurricanes Katrina and Rita would have been avoided had Louisiana not already lost 1.2 million acres of protective wetlands. The buffering effect of wetlands in storms exceeding Category 2 is not completely understood.[16] But nearly everyone agrees that the benefits are significant during storms both large and small. Inland wetlands like freshwater marshes, bogs, and more isolated "prairie potholes" also provide important flood control services, but to a lesser extent.

Experts estimate we have about 63 million hectares (156 million acres) of wetlands of all kinds left in the world. One recent report, cocommissioned by the Secretariat of the Ramsar Convention on Wetlands, attempts to quantify the "total economic value" of the world's wetlands by considering their various provisional, regulatory, cultural, and supporting services. Based on more than 200 wetlands valuation studies from around the world, the report puts the total economic value of the world's wetlands at around $3,300 per hectare per year, for a total of about $200 billion worth of annual services. Excluding cultural benefits, the most valuable single service, by a good margin, is flood control, which on average provides global benefits worth more than $450 per hectare per year. Numbers like these are far from exact. The flood control value, for instance, is based mainly on estimates of so-called indirect pricing methods, in which economists imagine a region without its existing wetlands and then estimate either the cost of the added flood damage ("avoided cost") or the cost of replacing the natural buffer with an artificial one ("replacement cost"). Sometimes important information is left out for logistical or other reasons. The Ramsar report, for instance, notes that its survey does not consider the value of "sediment control," among other things, resulting in what it calls an "underestimation" of overall wetlands values. Because of such factors, experts can and do disagree. The influential Costanza study, for instance, put the value of wetlands flood control services at $4,539 per hectare per year—10 times the Ramsar estimate. The Costanza study concluded that, in all, wetlands provide annual benefits worth $940 billion.[17] Such estimates, as I argue in part III, are too vague to shape policy but suggest the new respect ecosystems are earning among scientists and economists.

By any standards, wetlands are truly the world's meteorological speed bumps, slowing the ferocious stampede of hurricanes and typhoons on the coasts of dozens of countries. One of the most remarkable types of wetland barrier is the mangrove forest. These spidery trees flourish in brackish water on many warm coasts, often knitting themselves into thick bands along coastal shores and barrier islands. Their protective value is well known. For centuries, coastal fishers, merchants, and even pirates have slipped into their sheltered waterways in times of violent storms. Of the 177 countries in the world, about half have mangroves, including the United States (in

Florida only). The lion's share of this treasured resource (about 25 percent) belongs to Indonesia.[18]

Mangroves provide many different services. They help provide important consumable goods, such as fish and building materials. They support tourist opportunities and provide aesthetic and spiritual benefits to local residents. Mangroves also contribute to foundational ecosystem support processes by cycling nutrients and providing nursing habitat for young fish. Finally, mangroves are mighty wave-busters. Their multiple roots and stems dissipate the energy and size of waves by producing drag force. It is estimated that wave energy may be reduced by 75 percent in a wave's passage through 75 meters of water. One study suggests that in certain conditions, a 1.5-kilometer belt of mangroves could completely erase a wave 1 meter high. That service is worth money. In Malaysia, economists have priced the storm protection and flood control services of mangrove swamps at around $300,000 per kilometer, which is the cost of replacing swamps with rock walls. The storm protection benefits of mangroves in Thailand are estimated to be worth $3,679 in net present value per hectare.[19]

Unfortunately, the provisional services of mangroves have led to a devaluation of the others. It is estimated that 35 percent of the world's original mangrove cover is already gone. Some countries, like Myanmar (formerly known as Burma), have lost 85 percent, a fact that probably resulted in greater storm surges during Cyclone Nargis in 2008.[20]

When mangrove populations decline, it is typically because they have been converted to commercial fish ponds, sold as timber or firewood, used in charcoal production, or destroyed by disease and storm.[21] As a result, the coastal ecology changes dramatically as the habitat services that used to support coastal sea life erode. One irony is that fishers who destroy mangrove swamps to make way for industrial fish or shrimp ponds soon find their property—now choked with fertilizers and deprived of mangrove habitat— unsuitable for the very sea life they tried to enrich.

With the help of international aid, some countries are trying to restore mangroves for use as buffers and fish habitat. After the Bay of Bengal Cyclone in 1991, for instance, Bangladesh launched a program, called Coastal Green Belt, to line a third of the nation's shoreline with mangrove forests 2 kilometers wide. Already more than 120,000 hectares have been planted.[22] In the Thai Binh province of Vietnam, a Red Cross project helped plant 2,000 hectares of mangrove in shallow waters to protect a coastal dike system and to bolster the production of shrimp, crab, and other commercially valuable seafood.[23] Months before the project's planned evaluation, Thai Binh was struck by the worst cyclone in a decade. But the dike system, shielded by a flotilla of mangroves, remained intact.

Forests located farther inland also protect against storms by reducing the effects of rain-based floods and mud slides. In locations as diverse as the Himalaya, Brazil, and northern California, deforestation has allowed violent floodwaters to accelerate and level all kinds of development, from Third World shanties to cliff-side villas.[24] China's Yangtze River provides the setting for one cautionary tale, as reported by Janet Abramowitz of the Worldwatch Institute. In 1998 heavy rains in the river basin flooded 25 million hectares of cropland, caused $36 billion in property damage, forced millions of residents to evacuate, and took more than 4,000 lives. Summer storms are always a menace, but according to Abramowitz, "as the floodwater continued to rise, it became clear that other factors besides heavy rains were at play."[25] For one thing extensive logging and agriculture had left many steep hillsides bare. Indeed, 85 percent of the original forest cover in the river basin was gone. Gushes of mud rocketed down the downhill, carrying valuable crop land and even farm equipment along with it. Other natural flood control structures were similarly compromised. Dams and levees, for instance, prevented waters from draining into traditional floodplains. And many important wetlands and lakes had been filled in or drained, depriving the region of its natural capacity to absorb rain. The Chinese government, which initially blamed only natural factors, like El Niño, later appeared to acknowledge the role human development had played in sabotaging nature's protective infrastructure. The government banned logging in parts of the Yangtze basin and now pays former loggers to plant trees in the upper watershed. The government now estimates that the basin's regulatory value in controlling floods is at least 10 times greater than its timber value.[26]

In *Collapse: How Societies Choose to Fail or Succeed*, geographer Jared Diamond attributes much of the misery in Haiti to a failure of green infrastructure. Haiti and its neighbor, the Dominican Republic, are both located on the island of Hispaniola. But the two nations took different approaches to what was in the 1920s a lush and thickly forested landscape. The Dominican Republic managed to preserve much of its original forests. But because of many political and economic problems, Haiti's hillside forests have been nearly wiped out by logging. In 2002, Hurricane Jeanne barreled into the island, killing 1,800 Haitians, most of whom died in mud slides. In the

Dominican Republica, forests were able to reduce the flooding. Few deaths were reported in that country, with the exception of 400 people living in an exposed town near the Haitian border.[27]

Miles offshore, many coastal plains are blessed with chains of barrier islands that protect populated shores from storm-driven waves and tides. Such islands are thought to occupy 15,000 kilometers (9,320 miles), or 7 percent, of the planet's open shoreline. More than half of all barrier islands are located off the coasts of Europe, Asia, and North America, although they exist along every continent except Antarctica. Barrier islands generally lie parallel to the mainland, separated by estuaries, bays, or lagoons. Like lowland soils and wetlands, barrier islands exist in a dynamic environment. They are constantly being "shaped, reshaped, eroded, and accreted," their mass and form determined by local sand supply, wave energy, and tidal fluctuation.[28]

Scientists divide barrier islands into two categories, those on coastal plains and those along lowland river deltas. It is the second type that shields populous river communities from water and wind. And as it happens, deltaic islands, which make up a third of all barrier systems, are by far the less understood of the two.[29]

The most widely studied deltaic barrier islands are those protecting the mouth of the Mississippi.[30] These sandy humps, many christened with musical names like Chandeleurs and Timbaliers, help protect half a million people from violent storms, along with an international commercial-industrial complex worth billions. The seaward coast of such islands typically begins with a broad beach and sand dunes. The dunes contain vegetation that traps sand and helps stabilize soil. The sand, of course, has been provided by rivers and streams that transport sediment from hundreds or thousands of miles upstream. Behind the island dunes (toward the mainland) lie mudflats that are created when waves occasionally breach the dunes and soak into the ground. Saltwater marshes grow beyond the mudflats, providing important animal habitat and absorbent sinks in times of ocean surge.

In the Gulf of Mexico, deltaic barrier islands are under assault by many forces. Their proximity to sea life and picturesque views attract fishing communities and vacationers who build second homes on the bigger islands. Such development compromises the stability of fragile dune systems and marshland. Deltaic islands, which rely on constant supplies of rejuvenating sands, are starved of silt by rivers hemmed in by levees and dams. Rising sea levels, an effect of global warming, have also begun nipping at the heels of important barrier chains. Then there are the storms. The Chandeleurs, a

50-mile-long uninhabited chain of islands about 100 kilometers from New Orleans, has taken a beating over the last 10 years, including gut punches from hurricanes Georges (1998), Lili (2002), Ivan (2004), Dennis (2005), and Katrina (2005).[31] In fact, it is believed that Katrina, whose eye passed directly over the Chandeleurs, reduced the chain's surface area by half.

Or consider Isle Derniere, which in 1850 became the home of the first tourist resort on the Louisiana coast. It flourished for a few years with two nice hotels, but in August 1856 it was flattened by a hurricane that killed 140 people and destroyed every building except a hut. The island was never inhabited again. If you could flip through a series of small maps showing Louisiana's Isle Derniere as it existed at various times from 1850 to the present, you would see an animation beginning with a grinning slice of land reaching up at both ends, like the smile of a Cheshire Cat. The smile would curl and shiver for a generation or two, then slowly decay into twinkling wisps. Indeed, the appropriately named Isle Derniere (Last Island) is now more commonly referred to in the plural — Isles Dernieres — in reference to what few teeth the cat has left. Whether anything at all will survive the next generation is anybody's guess. According to the U.S. Geological Survey, Gulf island erosion is now occurring so rapidly that without restoration efforts, *all* of Louisiana's barrier islands could be gone by the end of this century.[32]

In the rest of the world, many important island chains are also in danger. In China, protective islands off the coast of the Po delta are literally falling apart, hammered by waves and robbed of sediment by upstream mining operations. In western Africa, island ridges on the Niger delta are eroding at a rate of 20 meters per year, endangering fragile mangrove forests that protect two of Africa's most densely populated regions. On the continent's opposite side, just north of the Nile delta, a 200-kilometer strip of islands is struggling to survive. This chain of barrier islands, one of the longest in the world, is seen by Egyptians as crucial protection for the delta's fertile lowlands. The islands are not only shrinking, but gradually *migrating* slowly toward the shore, a phenomenon that reduces their protective value. Both the shrinking and the migration are caused by sediment starvation. The same dams that prevent the Nile from replenishing delta lowlands with new soils also cut off the process of island building, called "accretion." Without new supplies of silt the islands erode, and tidal forces gradually sweep what is left into shoreline lagoons. The loss will only accelerate as other developing countries like Sudan and Ethiopia make good on plans to draw more water from the Nile River.[33] For Egyptians, the results could be tragic, eroding the delta, flooding hundreds of farms, and forcing the migration of thousands of residents.

The coral reef is among the world's most ancient types natural breakwater. Australia's Great Barrier Reef, said to be the largest single such structure in the world, is literally millions of years old. But even the relatively young barrier reefs off the coast of Florida were sprouting up some 6,000 years ago, long before Egypt's famous pyramids. But construction is slow going. On average, coral reefs grow about 1 to 16 feet every thousand years.

A reef's calcium-based skeleton is surprisingly durable and can significantly dampen sea surges before they reach the shore. Depending on a reef's width and underwater topography, coral reefs can absorb more than 90 percent of a wave's energy. In Sri Lanka, it is thought that every square kilometer of coral reef prevents 2,000 square meters of coastal erosion each year. Like the shielding effect of mangroves, the benefits of barrier reefs have long been appreciated. At the southeastern tip of India, for example, generations of villagers have assigned gender roles to the two parts of their coast. The more southern part of the coast, in the Gulf of Mannar, is said to be male because strong waves attack the coast but are blocked by an equally strong coral reef. The northern section of coast, which lacks coral, is said to be female—while the waters there are generally more subdued, when an angry storm comes, you had better look out. Fishing communities in the region still remember the 1964 cyclone that washed away one village but left another that was protected by reefs intact.[34]

Australia's Great Barrier Reef plays a similar if less dramatic role. By acting as a buffer against heavy seas, this complex of reefs and islands protects a shipping route hundreds of miles long and protects habitat for neighboring sea grass beds and coastal mangrove forests, which in turn protect Australia's coast from erosion and sea surge. Reefs of all kinds protect coasts in a less direct way, too, by producing sand (the product of calcified organisms) that forms and replenishes barrier islands and coastal soils. In addition to the regulatory service of shoreline protection, coral reefs also provide many consumable goods, including seafood, pharmaceuticals, building materials, and jewelry. Reefs also provide important cultural and recreational settings. The total annual economic value of coral reefs can be ballparked at somewhere between $100,000 and $600,000 per square kilometer. Recreational value often dominates estimates like these because tourism can be worth so much to a local economy. Still, the monetized storm protection value is significant. In highly populated regions of Indonesia, the barrier protection offered by reefs can be worth as much as $50,000 per kilometer per catastrophe based on the cost of replacing housing and roads that would otherwise be destroyed.

A study funded by the U.N. Environmental Programme grimly reports that "about 30 per cent of the world's reefs are seriously damaged, with possibly no pristine reefs at all remaining, and it has been predicted that 60 percent of reefs will be lost by 2030."[35] Things look worse in the world's storm belts. In Southeast Asia, 80 percent of reefs are said to be threatened by human activity, with half considered to be at "high" or "very high" risk. In the Caribbean, where human activity threatens two-thirds of all reefs, some experts predict that reef destruction will reduce shoreline storm protection by 10 to 20 percent over the next fifty years. Some human threats to coral have been known for a long time, including overfishing, "blast fishing," pollution, anchor damage, and tourism. A much more alarming but less understood threat comes from climate change. Higher ocean temperatures can kill sensitive populations of algae on which reef systems depend, leading to a phenomenon called "bleaching." On top of that, higher levels of carbon dioxide in the atmosphere will lead to higher carbon dioxide levels in the ocean, making the seas more acidic and incompatible with coral development.[36] Marine scientist Ove Hoegh-Guldberg concludes in his recent coral reef study in *Science* that "levels of CO_2 could become unsustainable for coral reefs in as little as five decades." He adds: "[W]armer and more acidic oceans . . . threaten to destroy coral reef ecosystems, exposing people to flooding, coastal erosion and the loss of food and income from reef-based fisheries and tourism. And this is happening just when many nations are hoping that these industries would allow them to alleviate their impoverished state."[37]

QUAKES AND SLIDES

The relationship between infrastructure and quakes and slides draws from some of the literature already discussed. Earthquakes on land present more hazards where soil is soft due to poor land-use patterns or filling of wetlands for development, a contributor to property damage from many quakes in the San Francisco Bay Area, including the 1906 earthquake and fire. In some cases earthquakes can be amplified or even triggered by human activity near fault lines. According to the U.S. Geological Survey, injection of fluids into deep wells for waste disposal has triggered earthquakes in Canada and the United States, including a moderate quake of magnitude 5.5 near the Rocky Mountain Arsenal outside Denver. Of greater concern, scientists in the United States and China are examining whether China's Zipingpu Dam may have helped trigger the 2008 Sichuan Earthquake (magnitude 7.9), which killed more than 70,000 people. The scientists think it is possible that the heavy waters of the dam's reservoir placed too much stress on the underlying rock. While the Chinese government—and some Chinese scientists—dismiss the idea, other scientists are urging a moratorium on new dam projects in China's earthquake-prone western regions.[38]

In contrast to earthquakes, landslides are more often the consequence of heavy rains and flooding. Thus many experiences with forests as buffers in the Himalaya, Haiti, and the northwestern United States also apply here. In the Pacific Northwest, which sees hundreds of landslides a year, a study found that 94 percent of them could be attributed to clear-cuts and logging roads. In 1996 alone, landslides cost the region billions of dollars in debris removal, reconstruction, and watershed restoration. Long ago the Swiss recognized the value of forests in protecting communities from landslides and avalanches. By some estimates, forests save Switzerland $2–3.5 billion per year. In the 2005 Pakistan earthquake, damaging landslides were much less frequent on forested slopes than on deforested ones, corroborating the experience of the Swiss.[39]

When volcanic activity or earthquakes occur under the sea, the force can produce mountainous waves called tsunamis that are capable of demolishing large metropolitan areas. Some scientists assume that the same features that shield coasts from storm surge also protect them from tsunami. That may be true in some cases, but the differences in wave amplitude and origin of force make general comparisons difficult. Studies of coastal damage in northern Sumatra after the 2004 Asian Tsunami, for instance, found that even flourishing barrier reefs did little to reduce onshore damage. The waves sailed right over the reef's top ridge, saving the coral but destroying the coast.[40] Analyses of satellite images taken over many areas damaged by tsunamis raise similar questions about the ability of mangrove swamps to prevent shoreline damage.

Still, there are some inspiring accounts of nature's counter-tsunami services. My favorite involves the village of Naluvedapathy in the southern Indian state of Tamil Nadu. Back in 2002, the community's few thousand residents decided to plant more than 80,000 trees outside their village, in a bid to enter the Guinness Book of World Records. In less than three years, they had transformed their land had into a miniforest of coconut palms and other tropical vegetation. When the 2004 Asian Tsunami roared onto the coast of southern India, many neighboring towns and villages were crushed. But Naluvedapathy, shielded by a kilometer-wide band of forest, remained nearly unscathed. In a BBC interview given shortly after this nearly miraculous event, a 70-year-old village grandmother named Marimathu showed she understood perfectly the relationship between disaster and natural infrastructure. Asked what advice she could give to fellow citizens on the Indian coast, Marimathu said this: "Tell others to plant trees."[41]

WILDFIRE

Like cyclones and earthquakes, wildfires are part of the natural world and help shape the ecosystems of which they are a part. Just as tropical storms might redistribute food to a growing fishery or help build a barrier island, wildfires play important roles in fertilizing soil, stimulating seed germination, and supporting wildlife.[42] But not all naturally occurring fires are desirable. Some endanger human life. And some reach an intensity much greater than a "healthy" natural system would normally allow. Indeed, wildfires in the United States have consistently grown in their intensity and destructive effects.[43]

One reason is the government's traditional policy of fire suppression on federal lands. The 1988 wildfires in Yellowstone National Park set off a debate about the appropriate role of forest fires on federal land. Before 1988, federal policy favored fire suppression in Yellowstone, which not only prevented the development of habitat as just described, but also allowed the accumulation of dense undergrowth, increasing the risk of larger, more explosive fires. Today, National Park policy is designed to "preserve the natural fire in the park" within reasonable limits. But fire suppression remains a widespread practice on other federal land, particularly where marketable timber is at stake.

A second cause of the increased intensity of fires involves the timber industry more directly. Logging disrupts the natural system of fire control by removing the largest and most fire-resistant trees from a forest. Taking the tallest trees not only removes a forest's natural "fire wall" but also allows more sunlight to reach the forest floor. More sunlight dries the younger trees and encourages the growth of more shrubs and bushes, all of which makes perfect fuel for forest fires. In the United States, analysis of the 2000 fire season—one of the most destructive in recent times—showed that the vast majority of fires occurred on previously logged and roaded areas.[44] In contrast, forests without roads and in wilderness areas saw relatively few fires if any. Because of their size and density, old-growth (also called primary) forests provide some of the best protection against fire, as well as floods, and landslides. Unfortunately, it is estimated that less than 5 percent of original old-growth forest still exists in the United States.[45] And we are quickly losing even that. According to a recent U.N. report, "[b]etween 2000 and 2005, the United States lost an average of 831 square miles . . . of 'primary forest.'"[46] The United States is logging its primary forests faster than any other developed country in the world.[47]

Things are even worse in countries like Indonesia and Brazil, where selective logging and clear-cuts occur on a much grander scale, driven by commercial greed, rural poverty, and lax regulation. The fragmented canopies allow surrounding trees and soil to dry out, particularly in times of drought. Thick rainforests that would otherwise rarely burn become super-sized tinderboxes. This is what happened in Indonesia's devastating 1997–1998 fire season that corresponded with the El Niña drought. Loggers and farmers set fires as a means of clearing land, but under conditions of drought and a thinning canopy, they quickly spread out of control. Nearly 10 million hectares—an area the size of South Korea—burned to the ground. The resulting smoke and haze spread throughout Southeast Asia, disrupting the lives of 70 million people. Shops and businesses closed. Schoolchildren were kept home. Many farms were incinerated, while others experienced indirect damage caused by haze and the disruption of pollination cycles. In all, Abramovitz reports, the economic loss was estimated conservatively to be $9.3 billion. "If harm to fisheries, biodiversity, orangutans, and long-term health were included," she writes, "the damage figure would be far higher."[48]

Aiding Recovery

While the protective capacity of nature is impressive, it's not foolproof. Living near the arctic, or on a tropical shore, or in earthquake country, means taking a chance—no matter how healthy the surrounding ecosystems. But ecosystems help build resilience in such areas that can allow communities to recover from naturally triggered disasters more fully and more quickly than they otherwise would. As reported by the World Conservation Union, "productive ecosystems can support sustainable income generating activities and are important assets for people and communities in the aftermath of disaster. For ecosystems to make these contributions, it is essential that they be factored into relief building efforts in the post-disaster response phase."[49] It is important to keep this second, and less obvious, role in mind as we consider resource preservation in terms of disaster management.

Consider this. During the New Orleans Flood, more than 30 billion gallons of water covered the city, trapped in an enormous soup bowl with no means of escape. The water sat there from three to five weeks before it was pumped out—mixing with toxic chemicals, decaying animal remains, and "hundreds of millions of gallons of sewage."[50] Where did that water go? The answer is Lake Pontchartrain, the massive estuary (it's not really a lake) on the city's northern border that empties into the Gulf of Mexico. Days after the flood, Army Corps officials began pumping Katrina's "septic soup" into the estuary. At first, environmental experts feared bacteria in the water would contaminate Lake Pontchartrain. (Indeed, the release, although necessary, probably violated the Clean Water Act.) But thankfully, the estuary, with a surface area of 630 square miles, was able to safely absorb the floodwaters and flush them into the Gulf. In fact, months later, shrimp catch in these waters actually rose, helping to prop up a battered fishing industry.[51]

Lake Pontchartrain is generally viewed as facilitating the New Orleans flood by allowing surge to enter the city's canals. But it was also an accomplice to recovery, providing a relatively safe "treatment" facility for billions of gallons of water and sludge. The estuary's resilience further aided the city by boosting fish catch and strengthening the state's hobbled economy. These fortunate events were not inevitable. Had Louisianans not made efforts to monitor and protect water quality in Lake Pontchartrain years earlier, the estuary's cleansing services could have been compromised. Had parts of it been filled or drained for development, its capacity for dilution would have changed. Lake Pontchartrain avoided these fates because residents cared enough about the commercial and recreational value of the estuary to keep it in relatively good shape. But by doing so, they got something extra—a "lagniappe," as locals say—in the form of water treatment services. And these services are not just for hurricane recovery. Lake Pontchartrain is also used to absorb water that is periodically diverted from the Mississippi River in times of severe flood risk. In 2008, for instance, Lake Pontchartrain swallowed millions of gallons of nutrient-laden water over several months to prevent levee topping in New Orleans and other communities upstream.

Stories like these must be distinguished from the patterns of ecosystem abuse that often accompany recovery efforts. In the developing world, it is well recognized that some of the worst ecosystem damage occurs not during the storm but afterward when recovery begins. "Not taking care of critical ecosystems after a major disaster," according to the World Conservation Union, "can cause significant economic and environmental losses, and impose hardships on already vulnerable communities."[52] Indeed, that organization notes that the 2004 Asian Tsunami was followed by a parade of recovery-based environmental harms, many of which impaired natural protections in the region. Evacuees were resettled in environmentally fragile areas, impeding human recovery and damaging habitat. Debris dumped into wetlands "blocked drainage, increased human disease and reduced the production of fish and other goods upon which local people depend."[53] Beach excavations brought in destructive invasive species. Meanwhile, residents desperate for building materials raided coastal forests (for lumber), beaches (for sand), and coral reefs (for cement). In Sri Lanka, in which many such events took place, it is estimated that misguided recovery efforts "caused more [environmental] damage than the tsunami itself."[54]

In New Orleans, mountains of contaminated debris were buried in landfills that were unsuited for such materials and prone to leakage. These landfills threatened the health of some working-class neighborhoods trying to rebuild. One landfill, less than a stone's throw from a national wildlife preserve, threatened to impair some of the very wetlands that protect the eastern part of the city from hurricanes.[55] As in Southeast Asia after the tsunami, some Louisiana marshes soon became illegal dumping grounds for everything from molding carpeting to abandoned refrigerators.

In Louisiana, even landscaping is now a political issue. Months after Katrina, as some less damaged neighborhoods began to recover, green thumbs were everywhere as people replanted their palms, hibiscus, banana trees, and oleanders. Of course they wanted the best mulch; and the best mulch, everyone believes, comes from hardy cypress trees. Where does this cypress mulch come from? Even the environmentally conscious landscaper I hired to replant my yard couldn't say. After inspecting the fine print on the several 40-pound bags then stacked in front of my house, I couldn't either. But in all likelihood, the cypress mulch in front of my house (which I sheepishly asked to be returned) and the cypress mulch on thousands of replanted yards throughout the city came from the very swamps that Louisiana depends on to anchor its coast and buffer its storms. How in the world do things like this happen? The answer has to do with the troubled relationship between economics and law. These topics are the focus of Chapters 3 and 4.

3

SYSTEM FAILURES AND FAIRNESS DEFICITS

What's more important, a forest or a shower? I thought about this 20 years ago when my wife and I were trekking in Nepal among the Annapurna mountains. One day, after ascending countless switchbacks in a cold rain, we arrived at a modest guest lodge just as the clouds began to clear. We had been pushing ourselves hard for a few days and were aching for a rest. Already the lodge (which was no more than a wooden hut with a few padlocked rooms) showed promise. The grounds were well kept and offered one of the most sweeping vistas of any place I had ever been. But that could not compare with what I saw next: a wooden stall beneath a tree with a hand-painted sign reading, "Hot Shower—30 Rupees." I arranged for the service, paid my fee (about a dollar back then), and hobbled into the shower stall. A sponge hung from a hook on the wall. On the floor (really a swept stone) sat two tall metal buckets, each filled to the brim with steaming hot water. You could say that without an overhead spigot, this was not a *real* shower. But the thought never crossed my mind as I reveled in what was surely one of the most luxurious experiences of my life.

But good fortune is sometimes followed by guilt. A few weeks later, I learned that hot showers like mine were an extravagance seldom enjoyed by local villagers. That's because heating up such quantities of water takes an awful lot of energy. At the time of my visit, 75 percent of the country's energy came from firewood, a statistic that hasn't changed much.[1] In addition, trekking-based tourism accounts for almost *half* of Nepal's base fuel consumption.[2] That means that travelers like me, whether hungry for hot meals or hot showers, indirectly consume a lot of forest material. To visitors on the trekking circuit, those forests still look healthy enough, with pines, rhododendrons, and junipers all crowding for attention. But off trail, the country is losing forests at a troubling rate. From 1979 to 1998, forest cover shrank from 43 percent to 29 percent, while the country's population increased by almost 60 percent.[3] That means fewer forest-based resources for local people, and less natural protection from the floods and landslides that have historically plagued the region.

This story illustrates several problems familiar to environmentalists. One problem involves a failure of the market. Markets are said to be efficient when the price of a product reflects its "true and total cost."[4] But when I paid a dollar for a hot shower—which includes the firewood that helped produce it—I was not paying for all the costs that might result from that lost tree. Those costs, among other things, include the loss of forest products, wildlife habitat, and protection against natural hazards. There are many reasons for the price discrepancy, including the dramatic wealth imbalance between rich and poor countries; but in this case the biggest factor may have been that lodge owners at the time of my visit were allowed by law to take firewood from public forests without paying for it. As a result, firewood consumed at a lodge was artificially cheap, and travelers like me consumed too much of it.

If you are wondering how lodge owners managed to get such a sweet deal from forestry officials, you have hit on a second, related problem—that of government favoritism and corruption. Lodge owners and other members of the tourist industry have long exercised broad influence over regulatory policy in Nepal. While tourism is important to Nepal, accounting for nearly 4 percent of its GDP and 18 percent of its foreign exchange, the industry's power should not go unchecked. Government officials often lack the resources they need to enforce the conservation laws that do exist, and in any event are predictably open to bribes, kickbacks, and other forms of baksheesh.[5]

A third problem, which might occur to one first, involves fairness. In Nepal, which is by any measure one of the poorest countries in the world, people depend on forests for the most basic provisions. In addition to fuel and soil stability, forests provide millions of rural Nepalese with food, medicine, shelter, shade, moisture retention, and animal fodder. Yet every year local people find it harder to live at subsistence levels because of a growing population combined with dwindling resources. While increased tourism and trekking revenue provide some (though not sufficient) offset, these benefits are hardly spread evenly across the population. Instead much of the profit is captured by lodge owners, the national government, and business interests in Kathmandu.

Chapters 1 and 2 introduced the idea of nature as infrastructure and documented the many ways that infrastructure in the United States and the rest

of the world is gradually disintegrating. Because natural structures like forests, reefs, and wetlands offer important protection against natural hazards,
reversing this devastation is vital to minimizing the risks of disaster. To do
that, we must understand why our laws have allowed valuable infrastructure
to be squandered so quickly. A government would never allow private parties to cannibalize the airports or dismantle the bridges for cheap steel. (We
often fail to *maintain* artificial infrastructure, but this is not the same as
destroying it.) Why is natural infrastructure treated differently? Why is the
Louisiana coast—under the watchful eye of the EPA, the Army Corps of
Engineers, various state agencies, and a pride of state and federal judges—
being destroyed at a rate of 30 square miles a year? Why is America losing
its old-growth forests faster than Bolivia or Sudan? Why are barrier islands
sinking in the mid-Atlantic and coral reefs dying off the coast of Florida?

The answers are suggested in the story of Nepal. Sometimes basic institutions like markets or governments fail to perform as they are supposed to.
When institutions fail, natural infrastructure is damaged or overconsumed.
Unfairnesses of many kinds then ensue. This chapter examines the failures
in markets and government regulation and links them to the special characteristics of natural infrastructure. I then describe how these failures result in
moral unfairness, whether in terms of human-centered values or ecology-
centered ones. While more abstract than earlier chapters, my analysis here
will provide a necessary foundation for assessing in concrete terms the failure of environmental policy in protecting natural infrastructure and for
imagining practical solutions—topics I will consider in Chapters 4 and 5.

Market Failure

According to advocates of green infrastructure, the market for ecosystem
services suffers from massive price distortion—an optical illusion in which
the benefits of resource extraction are made larger than life while the countervailing benefits of intact forests or marshland are minimized. Law professor Jim Salzman attributes the problem to three factors: ignorance, a narrow
view of economics, and lack of service-based markets.[6] While these problems were not intentionally created to mislead policy-makers, they are certainly taken advantage of by those with interests at stake.

IGNORANCE

Ignorance undercuts resource protection on two levels. The first level involves the public. We as consumers and citizens simply do not know enough

about the ecosystem services that support our way of life. When we see a television debate about logging, many of us might think of an ugly clear-cut or an imperiled owl, but not many will associate the loss of a forest with mud slides in an Oregon suburb or exploding wildfires in Montana. To pick another example, most people are aware that a lot of the seafood we eat is contaminated with mercury. But very few seem to know that mercury contamination enters the food cycle *from the air*, and that most of the airborne mercury comes from energy production, specifically coal-fired power plants. Until Hurricane Katrina, few people even in the Gulf states understood the crucial role cypress swamps play in protecting communities from storm surges.

These informational holes matter because citizens and consumers cannot make good decisions about environmental protection if they do not understand the risks at stake. Everyone knows the benefit of plentiful construction material, cheap electricity, and hardy mulch. But not everyone understands that the same things compromise many provisional and protective infrastructures that we take for granted. Recently, it seems, even well-educated jurists lack this basic information. In 2006 the Supreme Court reviewed a closely watched dispute involving the Army Corps's authority under the Clean Water Act to protect wetlands located near traditionally navigable waters.[7] Essential to the government's case was the argument that wetlands provide important services like absorbing and filtering floodwater. The idea was that if wetlands absorb the overflow of federally regulated waters, that connection should be enough to place those wetlands within the Corps's protective jurisdiction. Several friend-of-the-court briefs had been submitted by scientists and others describing how this or that marsh absorbed floodwaters and protected low-lying areas from harm. Such basic points are now common fare even in grade school classrooms around the country, proclaimed in posters festooned with green crêpe paper and pipe cleaners.

However, on the morning of oral argument, Justice Stephen Breyer seemed genuinely perplexed. Wouldn't *any* field next to an overflowing stream absorb water in the same way? he wanted to know.[8] When Solicitor General Paul Clement (representing the Army Corps) said no, and explained that wetlands have "unique characteristics" enabling them to act "something like a sponge," the justice appeared skeptical. That was a "scientific statement," Breyer pointed out, "requiring empirical verification." Referring to the Court's massive appellate record, he asked, "Where do I verify it?" Clement, who seemed abundantly prepared to answer any question but this one, vaguely directed the Court to its stack of expert briefs and agency documents, conceding that no government study may have actually used "the sponge word." Weeks after the argument, a law professor friend of mine joked that a simple Google search for the combination "wetlands" and "sponge" would have satisfied Breyer's curiosity. It was no joke: my most recent such search yielded 179,000 items, the first hundred of which backed the claim (after that I stopped checking).

The second level of ignorance is found at the technical level. No matter how smart or well-meaning our scientists and engineers are, they will always be pressed to do and explain more than the current state of knowledge comfortably allows. Chapters 1 and 2 have provided examples of this. Experts agree that the Louisiana marshlands lower the height of storm surge, but no one knows by exactly how much. We're impressed by the "Guinness Garden" of Naluvedapathy, which broke a tsunami's rage, along with a world record. But experts still debate the effectiveness of vegetation and other natural barriers in buffering quake-induced surge. As Abramowitz has pointed out, we also lack information about the size and health of even the ecosystems we know are vital.[9]

To make matters more complicated, many activities that harm natural infrastructure do so in incremental and cumulative ways, making their total harm difficult to predict. The oil and gas channels crisscrossing the Louisiana coast were not carved in a day but were dredged one by one—without, I suspect, much knowledge about the infrastructure loss that would come from any single project. The same goes for forests. No ecologist today doubts that clear-cutting a forest severely compromises services like water filtration and flood control. But today's timber projects, at least in the developed world, rarely involve total destruction of an ecosystem. As Salzman explains, "Much more common is marginal change—how will cutting twenty percent of *this* forest in *this* place impact water quality, flooding events, or local bird populations? In most cases involving a change in land use, whether it is forests, wetlands, or some other area, we simply do not know the answer."[10] Knowing the general ecosystem process is not the same thing as knowing the localized ecosystem service.

Because both experts and laypeople lack knowledge about the services ecosystems provide, it is extremely difficult to determine how much these services are worth to us. That's a problem because worth, or "price," tells markets how to allocate a resource. A high price leads to judicious use and perhaps conservation. A low price leads to gluttony. Since the days of Percy Viosca, Jr., economists have fashioned several ways to estimate the value of nonmonetary ecosystem services like water filtration or flood control. The method often involves calculating the costs of restoring a degraded service-providing ecosystem or of replacing it with something artificial. Sometimes economists will consider how an existing ecosystem "adds value" to a house or hotel room by increasing its market price. (On the Gulf shore, for instance, a beach house shielded by sand dunes is often worth more than one more exposed to the elements.) More controversially, economists sometimes use surveys asking people to hypothesize about the value they get from the aesthetic, spiritual, or contemplative experiences nature provides.[11] There are limitations to all of these methods, and figures sometimes differ dramatically. As we saw in Chapter 2, to use just one example, valuations of the world's wetland-based services vary by a factor of 10 (from $450 to $4,539 per hectare per year). Taken with a grain of salt, such numbers can help start a conversation about the need for infrastructure maintenance. But even with a price, the lack of service-based *markets* presents challenges for advocates of natural infrastructure.

A NARROW VIEW OF ECONOMICS

David Brower, a legendary environmentalist, had a favorite joke about capitalism as it applies to ecology. "Capitalism sounds like a great idea," he would say, with deadpan expression. "We should try it some time."[12] What Brower meant by this remark is that capitalism's high-minded ideas about efficiency and optimal use evaporate when discussion turns to resource allocation. One phenomenon he surely had in mind is the laundry list of "giveaways" bequeathed to special interests through historical precedent and pork barrel politics—federal mining and grazing rights at below-market rates, subsidized logging roads, and so on.[13] Informed economists do not even try to justify these actions on free-market grounds.

Brower's remark also suggests the failure of traditional economics to consider the hard costs society incurs when nature's infrastructure is destroyed. While it is easy to find markets for hot showers and Gulf oysters, there are no markets for the natural infrastructures that support these and other more important benefits. "As a result," writes Salzman, "there are no direct price mechanisms to signal the scarcity or degradation of [natural] goods until they fail (at which point their hidden value becomes obvious because of the costs to restore or replace them)."[14] In Canada, for instance, it is estimated that the average wetland spins off nearly $6,000 per hectare every year in ecosystem services when left intact, and generates less than half that when developed for intensive farming. In Thailand, healthy mangrove forests are thought to be worth $1,000 per hectare per year for their storm protection benefits and other services. But many are hacked down for shrimp farms that produce only a fraction of that value. And remember my craving for hot water in Nepal? According to the Millennium Ecosystem Assessment, "[i]n most countries, the marketed values of ecosystems associated with timber and fuelwood production are *less than one third* of the total economic value, including nonmarketed values such as carbon sequestration, watershed protection, and recreation."[15] Meanwhile, the logging industry chugs along unsustainably on six of seven continents.

The same problem exists on the global scale. The Millennium Ecosystem Assessment argues that a nation's inventory of renewable resources (including ecosystem services) and nonrenewable resources should be considered "capital assets" when measuring national wealth. "When estimates of the economic losses associated with the depletion of natural assets are factored into measurements of total wealth of nations," the report explains, "they significantly change the balance sheet of those countries with economies dependent on natural resources."[16] For instance in 2001, Ecuador, Venezuela, Ethiopia, and Kazakhstan, all showed positive growth in net savings, indicating a rise in national wealth. But factor in resource depletion and environmental damage, and net savings in those countries actually fall.[17] For international policy-makers who allocate investment aid on the basis of economic stewardship, these numbers make a difference.

One reason that ecosystem values remain invisible is that, as I discussed earlier, many are uncertain and unknown. Rather than pencil in a random number or draw a big "infinity" sign, traditional accounting measures presume the value is zero.[18] That practice is changing as ecological economics takes hold; but it is still the dominant practice in much decision-making. Another reason that natural services appear free is that we still indulge the hangover fantasy of unlimited resources. We know it's not true. But still it's hard to gaze down a trailhead, a river, or a frozen food aisle and admit that natural resources do not go on forever. Finally, as Salzman suggests, without price signals it is hard to know when nature is running on empty.

So traditional markets contribute to ecosystem loss by holding, in effect, that nature is worth more dead than alive. One reason this happens is that we do not have good ways of "pricing" the services that many ecosystems provide; and without a "price," we tend to undervalue ecosystem services. Here we are following a strong and widely shared cultural tradition of believing (against much current evidence) that nature is abundant and rejuvenating, or failing that, at least replaceable with technology. But this cannot be the whole story. If it were, data from the Millennium Ecosystem Assessment suggests that Canadians would not find it more profitable to fill in wetlands for farming. Thai landowners would not allow the destruction of their mangroves for shrimping. That is the puzzle. Even when we *know* that a healthy ecosystem provides more social benefit than an exploited one, and even when we *know* that its services are in danger of imminent and irreplaceable destruction—even in those clear cases—unregulated markets will encourage exploitation over preservation. Why?

The short answer is that because of the way markets are structured, the Canadian and Thai landowners are able to make good money through the sale of commodities produced on their land, but they are not able to make good money by leaving their property undeveloped. The property might be more valuable to their communities when left undeveloped and used as a water purifier or a storm buffer. But nobody's paying anybody for that, so the market wins and those services are compromised or lost.

The longer answer—which explains why the market is structured this way and suggests ways to change it—depends on a few principles of property theory. In *The Law and Policy of Ecosystem Services*, J.B. Ruhl, Steven Kraft, and Christopher Lant approach this problem helpfully by speaking of public and private property.[19] They divide the world into "private goods" and "public goods." (These terms, for simplicity's sake, also encompass services.) Private goods like cars or human labor show up in the market because of two characteristics: they are easily *excludable* on the supply side, and they are *rival* on the demand side. That they are excludable means that, for instance, if you want to buy my coral blue Prius, I can keep the keys until we agree on a price; or if you want me to write a legal brief for you, I can withhold my labor until we negotiate a fee. That they are rival means that, for instance, you cannot purchase my car or an hour of my time without depriving someone else of the same things: if you buy that hour from me, and my mother wants me to help fix her sink, she will have to wait.

Public goods differ from private goods in that they are harder to exclude and their use is mostly nonrival. New York City's Central Park is an example of a public good. It would be hard for the city to limit park use in any meaningful way without higher walls and more police officers. Even with those efforts, the city could not keep some New Yorkers from savoring the landscape, between sips of Chenin Blanc, from high-rise apartments on the Upper East Side. The enjoyment of Central Park is nonrival because your use of the park, whether walking the dog or enjoying the view from an apartment terrace, is not likely to detract from anyone else's use. There are always exceptions. In summer, Shakespeare in the Park gets crowded, and in December the Wollman ice-skating rink is a madhouse, but these temporary moments, which property theorists aptly call *congestion*, do not detract from the general idea. (A nonexclusive resource that is *regularly* prone to congestion is not a public good, and is sometimes referred to as a *common pool resource*.) There may, in fact, be no such thing as a perfect, or *absolute*, public good. Theorists used to point to the ambient air and outer space as purely public things. But air is often congested with pollutants that harm others, and Earth's orbital space is so cluttered with gear that the U.S. government now electronically tracks tens of thousands of artificial objects (from satellites to lost wrenches) so as to distinguish them from incoming missiles.[20]

Infrastructure, in general, bears the characteristics of a public good. The highway, the bridge, the broadband network—whether or not publicly owned—are meant to provide benefits to all users (nonexclusivity) and aspire to be so plentiful as to avoid routine congestion (nonrivalry) Essential to the concept of infrastructure, and to many other public goods, is an extra spillover effect that accrues to the public at-large even after individual users have had their share. Chicago residents may fly a little or lot, but they all benefit from a world-class airport nearby. A soft drink vendor might never see a free Shakespeare in the Park performance, but might still profit when the bard plays in the Big Apple.

In describing the public benefits of infrastructure, economist Edward Steinmueller uses the word *synergy*:

> [U]ses of the term infrastructure are related to "synergies," what economists call positive externalities, that are incompletely appropriated by the suppliers of goods and services within an economic system. The traditional idea of infrastructure was derived from the observation that the private gains from the construction and extension of transportation and communication networks, while very large, were also accompanied by additional large social gains. . . . Over the past century, publicly regulated and promoted

investments in these types of infrastructure have been so large, and the resulting spread of competing transportation and communications modalities have become so pervasive, that they have come to be taken as a defining characteristic of industrialized nations.[21]

It's hard to charge residents for "synergies," let alone other benefits that are widely shared among multiple groups. This is why national and local governments often bear much of the expense for artificial infrastructure like sports arenas and bridges. When artificial infrastructure is owned or maintained by private parties, there is almost always a way for the private party to profit by taking advantage of a "bottleneck" in the flow of a service. The bottleneck on a bridge is the tollbooth, which must be passed through to enjoy the benefit. For a landfill, it's the truck entrance where a tipping fee is charged. Many kinds of natural infrastructure, particularly those related to disaster protection, lack that kind of bottleneck. This makes it harder to link users of infrastructure with providers of infrastructure to produce some kind of "sale." This brings us to Salzman's third complaint about traditional economics as applied to ecosystem services: that unless we find new ways to link users and providers, markets will do a bad job of encouraging even the kinds of resource preservation that all would agree make society better off.

LACK OF SERVICE-BASED MARKETS

According to Salzman, "Markets for ecosystem services can only be established if there are discrete groups of buyers (service beneficiaries) and sellers (service providers). Otherwise, transaction costs become too high for contract formation."[22] He uses the example of biodiversity—a region's variation in life forms, which produces enormous benefits for agriculture (by providing variations in crop characteristics) and medicine (by providing antibiotics and other medicinal compounds). We lack a true market for this treasure because "there is no sufficiently discrete class of beneficiaries with whom we can negotiate, and the transaction costs of gathering enough beneficiaries together to negotiate for the service are too high."[23] Another important reason for this defecit is that our lack of knowledge about ecosystem services is related to their invisibility in the marketplace.[24] When there is money at stake, people will spend a lot to learn about, say, the loft of goose down or the roundness of tomatoes. Yet, as was shown in Chapter 2, experts know much less about the role of coral reefs during a tsunami or the best way to restore a cypress swamp. Salzman does not miss this point. Our lack of appreciation for natural flood control, he writes, "was tragically evident in the recent flooding in New Orleans. The wetlands that could have slowed the floodwaters were steadily degraded over time. . . . As floodwaters rose in New Orleans, people realized the importance of services that could have been provided by wetlands, but this recognition was too little and too late."[25]

To give this critique more context, let's look at a not-so-hypothetical situation. I know a gentleman who owns thousands of acres of swamp and marsh in the Atchafalaya Basin, about 100 miles southwest of New Orleans. It is gorgeous land and productive in many of the ways that Viosca and Lockwood would have recognized. My acquaintance is a bright, caring soul. He is, in fact, a descendant of the Isleño immigrants for which Bayou des Familles was named. He grew up fishing and hunting in the Atchafalaya and loves the outdoors. So what does my friend do with his many acres of wetlands? Savvy in business, he knows he has many options. He could lease his land to recreational groups who take visitors on motorboat rides through the cypress. He could lease his land to duck hunters. He could lease his land for use as a "'gator farm," where newly hatched alligators would be brought to lounge about and grow fat until they were old enough to capture and skin. Or he could go for the big money and lease his land to oil and gas companies, which would erect wells, dredge channels, and lay underwater pipe. In fact, my landowning friend has done *all* of these things over the years and appears to have profited from these endeavors, despite widespread erosion caused by the oil and gas people.

What my friend has *not* done in these many years is protect his land from erosion so as to shield his upstream neighbors from hurricanes. Remember, he's a caring guy. But he knows he can make a lot more money leasing his land to an extractive industry than he can by leasing his land to his upstream neighbors. That is not because the buffering services his land provides have no value but because those public services are harder for him to market. One reason for this is that he would not know where to direct his offer. Because of knowledge gaps in science, we don't know exactly which communities would be protected by these wetlands in any given storm; and besides, storms are all different. Another reason is that because storm buffering is nonexclusive and nonrival, market forces will not assign a high enough price for the service. A landowner can't, after all, instruct his swamps to shield only the paying customers and let the rest drown. Similarly, the land-as-shield transaction presents the ultimate free rider problem. Once one neighbor has paid the landowner to preserve buffer land, no one else is going to make a second offer because the first neighbor's "purchase" does not prevent the second neighbor from benefiting for free.

All of this leaves us with a dizzying market failure. To review: ignorance prevents us from knowing the value and importance of natural services. Traditional markets do not reflect the value, because ecosystem services often take the form of public goods, whose benefits are hard to sell directly to members of the community. Without a way to sell the benefits, the owner of an ecosystem has no motive to preserve its services or even learn more about their value, economic or otherwise.

One can imagine several ways to correct market flaws like these, from specialized private markets to aggressive public regulation. Advocates of ecosystem services seem especially enthusiastic about the market-based solutions, particularly ones in which environmentally conscious resource owners are compensated directly or indirectly by armies of thankful resource users. Biologist Gretchen Daily, a leader in the field of ecosystem services, would experiment with more private ownership of natural assets. She imagines "a place where people can get together to bargain — a market, whether in the town square or on the Internet" — in which asset owners and asset users could hammer out conservation agreements.[26] Salzman urges readers to be open-minded about the suggestion that government pay resource owners not to degrade their land.[27] Because of these and other talented thinkers, my mind remains open. But we should remember that while it is tempting to remain neutral on the question of who pays for conservation (as between the owner and the user), considerations of community capability and distributional fairness might sometimes tip the scale. In addition, as experts like Daily are careful to emphasize, any approach to asset protection, market friendly or otherwise, will require a strong and comprehensive regulatory role for government.[28] This brings us to the subject of good governance.

Government Failures

The evil twin of market failure is a phenomenon some political theorists call "government failure." The term describes situations in which publicly beneficial outcomes are thwarted by imperfections in government process. We've seen how optimal social gain through private markets can be undercut by things like ignorance, externalities, and collective action problems.

In government, many of these same elements can conspire to kneecap the public good. Legislators and regulatory officials are supposed to decide policy on the basis of "the needs of the American people." But often they don't. According to an economic model of politics called "public choice theory," this is because officials have only a weak incentive to take care of public assets because the assets they save and the resources they expend are not their own. That is, the positive and negative outcomes are externalized to others. One might expect politicians to work tirelessly on the public's behalf in the hope of pleasing voters and getting reelected. But public choice theorists downplay this incentive. They believe that voters are generally ineffective at monitoring their public leaders and holding them accountable. This is because most voters are uninformed (ignorance) and, except in the case of blockbuster issues, lack incentives to organize politically (the collective action problem).

On the other hand, regulated businesses and single-issue lobby groups are much better at influencing politicians. Unlike you or me, they have a focused interest in, say, a new mining lease or a proposed shipping canal; and so they are more willing to concentrate their resources, including large political donations, to change political minds. As for officials at federal agencies, they are also prone to special interest lobbying, particularly if they have previously worked in the industries they now regulate. Because the public follows regulatory matters even less than it does legislative ones (and because they don't vote for agency officials anyway) agency accountability to real people is not insured. This phenomenon appears to have been at work in the situation of Nepal and its forests. Lodge owners and other players in the trekking business were able to join forces and push for their economic interests in a way that Nepal's rural populace, which is spread out geographically and often poorly educated, was not able to do. To make matters worse, many of the things the tourism lobby wanted didn't cost much in market terms. For the most part, it wanted wider access to public forests and other natural assets—requests that would not threaten an existing agency budget or require a politician to shift funds from one constituency to another.[29]

The price of government failure is perhaps most obvious when studying artificial infrastructure, where monetized costs and benefits are easier to tally than in the natural world. In a recent piece in the *Atlantic Monthly*, Bruce Katz and Robert Puentes of the Brookings Institution investigated U.S. spending on transportation infrastructure.[30] Transportation infrastructure is, of course vital to economic output, producing discrete economic inputs for certain industries as well as more general synergies. Katz and Puentes found that the bulk of this economic output (75 percent) is generated in the nation's 100 largest metropolitan areas. These areas handle 75 percent of foreign sea cargo, 79 percent of air cargo, and 92 percent of air passenger traffic. One might expect federal transportation spending to emphasize these areas, but Katz and Puentes find that it doesn't. Instead transportation spending, which in 2007 included $50 billion for surface transportation alone, is "spread around the United States like peanut butter."[31] The reason has to do with earmarks, the provisions that members of Congress routinely put in legislation to direct public funds to specific projects. Earmarks are seen by many as gifts meant to impress local voters or generous campaign donors. Because they are rarely researched or debated, earmarks often have nothing to do with economic utility or "the needs of the American people."

In the past, Katz and Puentes report, strategic investments in railroads and highways "turbocharged" the economy and changed the American lifestyle. As recently as the 1970s, public investment in transportation infrastructure generated a return of nearly 20 percent. But since then, the popularity of congressional earmarks has grown, and government return on investment has plunged. Transportation investments in the 1980s "generated only a 5 percent return," they write; "in the 1990s, the return was just 1 percent."

As with market failure, legal scholars and others have proposed many ways to address the inefficiencies of government failure. Some would cede more decision-making power to private markets, reasoning that the kinks in a market system are less harmful or easier to straighten out than the python coils of special interest politics. That approach often results in calls for deregulation, small government, and other shibboleths of neoliberal economics. Others would make government behavior less arbitrary and more rational by requiring officials to follow more formalized decision-making methods (for instance, cost-benefit analysis) or by subjecting officials to legislative oversight or public scrutiny. For now, it is important to remember that as with market failure, solutions should be practical and need not follow a single ideology. In addition, our choices should in part be guided by a desire to enhance social fairness.

Fairness Deficits

When I tell students about my hot shower in Nepal, their first reaction is generally one of concern for that nation's struggling villagers whose essential resources are quickly dwindling. They might also mourn the ecological collapse of a region so rich in religious and cultural importance. It is only after some prodding that my students begin to see the problem as a failure of the market or of government. (To my regret, it's only at this point that most students start taking notes.) Most of this chapter has focused on the dynamics of markets and politics because these ideas perhaps best help explain *why* natural infrastructure, such as the forests of Nepal or the marshes of the Atchafalaya, is allowed to deteriorate. Concepts of public and private goods also help explain why unregulated markets cannot be expected to "just fix themselves."

The language of property and economics also helps us locate an important reason why this state of events is undesirable. When important natural infrastructure is destroyed or misused, social utility declines. We are all worse off because the positive externalities we would otherwise enjoy are lost. The countervailing benefits of destruction (which go to a smaller group) are not enough to make up the difference. The role of "synergistic" positive externalities in promoting social utility is well known and is a main justification for maintaining other kinds of infrastructure as well, such as sidewalks, roads, and bridges.

But as I pointed out in Chapter 2, access to infrastructure should also be just. It is that point my students see before any other in our discussion of Nepal. To many of them, it is simply unfair that resources important to a people's very survival should be consumed by a smaller, more fortunate group (tourists) to satisfy less essential needs. This discussion has many threads but comes down to two objections. The first is that this deprivation of flood protection, nutrients, and other services intrudes on people's *liberty* by denying them the basic ingredients of a safe and dignified life. The second is that this deprivation also violates a principle of *equality* by allocating resources in a way that favors the few over the many.

Environmental policy, or any policy for that matter, is about more than social utility. Fairness deserves equal billing. In the real world, most people, including judges and legislators, give fairness independent weight in policy analysis. Even traditional law-and-economics scholars generally admit that social utility by itself is an insufficient guide to moral governance.[32] Joseph

Singer, a professor at Harvard Law School, has spent his career writing about fairness as it relates to property. Many of his insights derive from what he calls the "social relations approach" in property theory—the idea that property rules are not about places or things but about human relationships. In the more traditional view, Singer notes, property is associated with "the idea of autonomy within boundaries."[33] According to that idea, the question whether my landowner friend in the Atchafalaya can carve up his marsh at the expense of upland neighbors is a matter of what control he should have within the boundaries of his own land. In contrast, Singer asserts that true autonomy derives not from "isolation" but from "relationship from others."[34] In this perspective, "people are situated in a complicated network of relationships with others, from relations among strangers, to relations among neighbors, to continuing relations in the market, to intimate relations in the family."[35] In order to understand autonomy in this richer sense, writes Singer, "one must discuss the effects law has on human relationships and the effects human relationships have on law. More fundamentally, one must determine which relationships to foster and which to suppress."[36]

Applying Singer's idea to resource protection, we can say that those who control natural infrastructure (whether through ownership or government acquiescence) are in a relationship with outsiders who rely on that infrastructure for important services. Nepalese villagers are in a relationship with lodge owners who legally (and sometimes illegally) mine the forests for fuel and sell it for profit. New Orleanians are in a relationship with the shareholders of oil and gas companies, whose actions legally (and sometimes illegally) have destroyed great swaths of protective wetlands, exposing hundreds of thousands of people to unnecessary danger. To say that the lodge owners have a license to take public firewood or that oil companies have a lease that permits dredging is only the beginning of the analysis. A property interest under the social relations approach is never absolute and must be partly defined by the important needs of outsiders who reasonably depend on aspects of the resource at issue.

In relationships involving property, it is natural to speak of liberty interests. Most of the time we have in mind what philosophers call "negative liberty," that is, the freedom from external restraint. It is negative liberty that protects protect property owners from unreasonable intrusions by government, meddling strangers, or even other property owners. The social relations approach acknowledges this but also recognizes the importance of "positive liberty," the ability to pursue one's own objectives.[37] Positive liberty may be seen as the rationale behind health and safety regulations or consumer protection laws. The idea is that every citizen should have access to some minimum level of resources so as to allow the pursuit of a safe, purposeful, and dignified existence. This approach helps explain, for instance, why air pollution regulations in the United States have for many years kept factory owners from polluting air above levels that might endanger public health, without regard to financial cost (although cost is considered indirectly through a state's implementation of the rule). The proximity of the factory to the public puts the factory owner in a relationship with her neighbors. And her neighbors' health, so essential to the exercise of liberty, may justly be deemed more important than the owner's costs or even society's overall utility. Internationally, a framework for measuring positive liberty, pioneered by Martha Nussbaum of the University of Chicago, has been adopted by the U.N. Development Programme for use in its human rights reports. This framework, which Nussbaum calls the "capabilities approach," pays particular attention to access to environmental resources and is viewed as especially important in assessing the welfare of women in developing countries.[38]

The social relations approach also emphasizes equality. Just as a property owner may be limited by society's interest in promoting liberty for others, he may also be limited by society's interest in promoting equality for others. The often-cited example of public accommodations law makes the point. Although the right to exclude has traditionally been seen as fundamental to private ownership, American courts today have no problem upholding laws that prevent shop owners from excluding customers on the basis of their race, sex, or other personal characteristics.[39]

Incorporating fairness expands our analysis of natural infrastructure by adding some elements that one does not often see in discussions of ecological economics. First, the rhetoric of fairness reminds us that there is a reason beyond social utility for protecting ecosystem services, particularly protective infrastructure, on which people's lives depend. That reminder is helpful for policy-makers looking for additional justifications to protect an old-growth forest or a chain of barrier islands. But it makes things more complicated in situations where the interests of utility and fairness diverge. As I will show in Chapters 4 and 5, policy-makers are already confronting dilemmas like this. Rather than ignoring fairness concerns, I believe insights from the social relations approach can point the way toward compromise.

Second, fairness provides another indicator for valuing the social importance of an ecosystem service. Straight resource economics relies on social utility to give these services priority. As we have seen, that utility is often expressed as a monetized sum—$1,000 every year for a hectare of Thai mangroves, or $6,000 per year for a hectare of Canadian wetlands, to use examples we've seen before. But these numbers say nothing directly about these ecosystems in terms of their ability to promote liberty or equality.

It is possible, of course, to argue that monetized values *indirectly* reflect fairness concerns by referring to markets in which consumers have already expressed their preference for fairness in the prices they are willing to pay.[40] But that seems like a dodge, since it is doubtful that people consciously incorporate their preference for fairness in consumer decisions. And even if they did, their willingness to pay would be affected by their ability to afford, granting the preferences of the wealthy more weight than the preferences of the poor.

Finally, a concern for fairness could in some cases suggest which party should pay the cost of restoration or preservation. As we've seen, resource economics is conceptually silent on this question. Most scholars in this field seem motivated by practicality as much as anything else. Their interest in creating new markets in which users or government agencies pay landowners to preserve natural services appears to be a reaction to an antiregulatory bias in many rich countries and regulatory impotence in many poor countries. In contrast, fairness takes a normative stand against the hoarding of resources that are necessary to liberty and equality. Political realities cannot be avoided, and resource protection will always be a process of compromise. But in the abstract, the question of who pays should not be determined by a coin flip. Fairness should be considered and used as a basis for negotiation.

4

TENDING OUR GARDENS

In Chapter 1 we toured Louisiana's Barataria Basin and began thinking of nature as an elaborate infrastructure, a source of interrelated goods and services that are impossible to replace. In Chapter 2 we pulled the lens back to examine natural infrastructures around the world, paying special attention to those that protect human communities from natural hazards like storms and landslides. What we saw, in many cases, were important ecosystems withering as if on life support. Lush forests that screen fires and absorb flood waters are disappearing. Barrier islands and surge-blocking wetlands are sinking into the sea. That loss has great significance in terms of both utility and basic ideas of fairness. Open markets and democratic governance are supposed to enhance utility and fairness. Chapter 3 reminded us that many nations (Nepal being one example) have underdeveloped economic and political systems, a fact that contributes to resource depletion. Even democracies with robust markets, like the United States, fail to use natural infrastructure in ways that are efficient, sustainable, and fair. That's because failures in markets cause many services provided by natural infrastructure to be discounted and degraded over time. Failures in governance—special-interest politics and corruption—lead to misallocations in infrastructure investment, in terms of both artificial infrastructure *and* natural infrastructure.

It is the law's job to minimize these failures and to make sure that natural resources are used justly and wisely. That is particularly true in such life-and-death matters as disaster management. In the United States, law is increasingly falling behind in this mission. Decades ago, America was the undisputed leader in environmental protection. Our laws and ideas were studied and copied around the world. Today we're coasting in the "sag wagon," with much to learn from the European Union, Japan, and others.[1] If we fail to reinvigorate our laws and enforcement methods, we are doomed to lose the irreplaceable infrastructures that keep many of our communities and commercial activities alive.

The way to get started is to survey the kinds of laws we use to plug up the holes one commonly finds in market and government processes, and to think about how the values of natural infrastructure can be highlighted. We can usefully sort types of environmental protection strategies into four categories: public management, public regulation, economic incentives, and Property Interests.[2] Public management includes all the ways government manages the natural resources it *owns* or *controls* for the benefit of the public at large. The management of public forests is a good example of this, as is the administration of navigable rivers and many coastal resources. Public regulation includes laws that control the behavior of owners and operators whose activities affect the environment; regulations like these control the activities of private actors as well as public agencies. Economic incentives include taxes (or fees) and government subsidies. The category of property interests encompasses an array of rights or entitlements held by citizens as individuals or as representatives of the public. These interests derive from common law (as in the theory of nuisance), Roman law (as in the public trust doctrine), and constitutional law (as in the protection against government takings), or some combination. A clever sorter could probably fit these ideas in the three previous categories, but I keep this class separate because it aids my analysis.[3] (The property interest category also includes "cap-and-trade" regimes like those that regulate sulfur-dioxide emissions in the United States and greenhouse gas emissions under the Kyoto Protocol; but I will not discuss cap-and-trade regimes here.)

If we were designing environmental policy that emphasized protection against disaster, we would ask it to do three things: protect the natural functions that minimize hazard, encourage human development to take full advantage of those functions, and respect the inevitable limits of natural functions. Let's take a look at where American law stands with regard to this template by considering some instructive examples.

Public Management

Infrastructure, whether natural or artificial, is often an object of public stewardship. The idea is that because the benefits are so diffuse and the user base so broad, it would be impractical to charge individuals for use of a dam or a long shoreline. In addition, the use of such things (to store drinking water, access the sea) are often seen as so basic to human welfare as to be entitlements, not subject to one's ability to pay.

The largest landowner in the United States is, of course, the federal government, managing 650 million acres throughout the 50 states (in addition to millions more owned by state and local governments).[4] On paper, federal laws and policy recognize our natural heritage as a source of wealth, security, and inspiration. Statutes governing publicly-held lands often speak of sustainability and the need to protect important resources for future generations. But the dream is not the reality. Law professor Alyson Flournoy notes that over the years the nation's forests, grasslands, parks, and wild lands have been steadily degrading. Her recent survey of public resource management found "a systemic pattern of squandering public resources or failing to take necessary steps to protect them, notwithstanding stated commitments and mandates under existing law to use them sustainably."[5] Thick forests, which help contain floods and wildfires, are being imprudently logged, sometimes without environmental review. Range- and grasslands, which serve as absorbent floodplains, are allowed to erode. Oil and gas development, which can cause underground fissures and dangerous subsidence, is unwisely allowed in vital marshes.[6]

The problem is that there is no clear mandate for protection or for considering the value of ecosystem services, including the services of protective infrastructure. Without any clear mandate, commercial or industrial lobbies are able to "capture" the system, tilting public management toward profitable, unsustainable use. (To review: the short-term profit motive of these lobbies is sustained by the market failure discussed earlier—that is, a market that fails to explicitly recognize the long-term benefits of preserving natural services; these lobbies are able to bring policy-makers to their way of thinking by concentrating their lobbying efforts in ways most citizens cannot, thus contributing to government failure.) This situation is even worse for natural than for artificial infrastructure. Everyone agrees that a bridge is to be used as a means for vehicles to cross a river or gorge. There are few "multiple uses." One can imagine a debate between the needs of vehicles versus bikers, or cars versus trucks; but there are no debates about whether a bridge should be used now or saved until later. Or whether its use as an object of contemplation or as a hedge against disaster renders it inconsistent with its use for transportation. To see this in practice, let's look a two examples, forest policy and water management.

FOREST MANAGEMENT

Most federal land is managed under a policy framework called "multiple use/sustained yield," (which I'll call simply "multiple use"). More than

192 million acres of national forests, 83 million acres of national parks, and 260 million acres of land administered by the Bureau of Land Management fall under this form of management.[7] While the language varies, the U.S. Forest Service's multiple use policy captures the idea: Congress charges the Forest Service with "the management of all the various renewable surface resources of the national forests so that they are utilized in combination that will best meet the needs of the American people; making the most judicious use of the land for some or all these resources or related services over areas large enough to provide sufficient latitude for periodic adjustments in use to conform to changing needs and conditions."[8]

This is hardly specific. Indeed, courts have found that this directive allows enormous latitude in determining what best meets "the needs of the American people." Under this model, the Forest Service need not use any particular accounting practices or even consider natural infrastructure services in their calculations, so long as they characterize the data as uncertain.[9] While this discretion could allow for aggressive protection of natural infrastructure, federal agencies have refused the opportunity. In addition, the aggressive trend toward cost-benefit analysis in regulatory review has, ironically, buried ecosystem services even deeper underground because they are so hard to monetize. So much discretion has left regulatory agencies vulnerable to industry lobbyists who profit little from protecting natural infrastructure for others. For these reasons, law professor Ruhl, Kraft, and Lant finds that the federal track record of using ecosystem service values to inform land use decisions has been nothing less than "repugnant."[10]

Forest policy during George W. Bush's administration was especially dismissive of ecosystem service values and the scientific information on which they are based. In 2003, for instance, he signed the Health Forests Restoration Act, which was said to be intended to reduce the risk of wildfires by making it easier for timber companies to log in overcrowded forests, thus removing "hazardous fuels."[11] But on closer examination the legislation mainly paved the way for more logging, doing little to minimize fire risk. For instance, most experts emphasize the need for localized study in order to draw any meaningful conclusions about fire in a forest ecosystem; but the Healthy Forest Restoration Act made forest thinning a blanket policy and weakened requirements for site-specific environmental analysis.[12]

Such a policy is not supported by the science. As Flournoy writes, "research documenting the effectiveness of broad-scale fuel reduction treatments for reducing the extent and severity of wildfires is generally lacking."[13]

Even worse, indiscriminate thinning could lead to *more* fires, not fewer. That's because logging itself increases fire intensity by leaving behind much easily combustible debris. Some experts argue that the Act's loose standards would in some cases allow timber companies to remove even the "large fire-resistant" trees, again defeating the law's primary purpose.[14] These inconsistencies have led many environmentalists to say that the Act was really not about fighting fires at all but was instead a sly attempt to give cover to the buzz-saw crowd. Whether or not that is true, the Act's ineffectiveness in protecting forests highlights the need for local scientific study when restoring natural infrastructure so as to minimize disaster.

Near the end of the Clinton administration the Forest Service proposed taking the remaining roadless areas, 60 million acres of some of the most nearly pristine wilderness left in the forest system, and making them off limits to roads and logging. The Forest Service argued that its so-called roadless rule would protect not only these ancient forests but also the many services they provide, which as we have seen include soil stabilization, fire control, and flood protection. The rule was a hit with the public, but not with the timber industry—nor, as it turned out, with the Bush administration when it came into power. Although the rule remained federal policy, the administration refused to defend it against legal challenge from the timber industry. But more troubling to advocates of green infrastructure was the Bush administration's use of regulatory review to attack the rule. In a 2002 report to Congress, the White House Office of Management and Budget (OMB), the federal agency charged with evaluating the efficiency of federal proposals declared the roadless rule a financial bust. In its estimation, the quantifiable cost in terms of lost jobs and foregone logging totaled more than $180 million per year, while the quantifiable benefits barely reached $219,000 per year. Those benefits, by the way, came not from mitigated floods or mud slides—which presumably fell into a black-hole category called "nonquantifiable"—but rather from money saved by not having to build and maintain the roads in the first place.[15] Without the administration's support, national forest policy devolved into a patchwork of regional practices and state rules, clearing the way for logging projects in some states. In the spring of 2009, the Obama administration issued a one-year moratorium on commercial activity in roadless areas to allow time for a permanent replacement rule to be crafted.

The Bush administration also seriously undercut the government's "forest management plans," which designate how forested lands will be used, what areas will be open to logging, and how ecological health and services (including flood control) will be monitored. In 2000, the last year of the Clinton administration, the Forest Service issued new regulations for forest management plans, completing a revision process that had been in the works, off and on, since 1982. Among other provisions, the new rules put more emphasis on sustaining ecosystem services and increased avenues for public participation and required that decisions be based on the "best available science." The Bush administration, displeased with this development, delayed implementation of these Clinton-era rules for nearly five years before finally replacing them with a new set of regulations in 2005 that emphasized economic sustainability over ecological sustainability, decreased public participation, and weakened the role science and site-based monitoring in making decisions.[16] The new regulations, which in some ways were less protective than even the 1982 regulations of the Reagan administration, eventually drowned in a whirlpool of litigation. In 2007, a federal district court ruled that the Forest Service, in proposing and issuing the 2005 rules, had managed to violate the National Environmental Policy Act (NEPA), the Endangered Species Act, and the Administrative Procedures Act. The court enjoined implementation until the agency "fully complied with the pertinent statutes."[17] A revised planning rule was issued in 2008, but that rule too was rejected by the court.[18] As this book goes to press, the Obama administration is in the course of developing a new proposed rule.

These brief examples show an overall need for greater attention to natural infrastructure, buttressed by the best science and strong public participation. Good science helps policy-makers understand the nonmonetary values of natural services (which otherwise fall off the balance sheet) and develop ways to maintain or restore them. Strong public participation ensures that decision-makers will hear not only from the industries that profit from a forest's commodification but also from members of the public and the scientific community who value the forest's other benefits.

More specific reform efforts should begin with a full scientific inventory of forest-based ecosystem services, something like the Millennium Ecosystem Assessment examined in Chapter 2. The inventory would document the role of forests in flood control, fire suppression, and land stabilization, all services that are important to disaster mitigation. The analysis would pay particular attention to geographic location and surrounding development. This information could then be used to inform future forest management plans and NEPA reviews. Thus changes in those areas should also be contemplated. From 2003 to 2005, the Forest Service invoked categorical exclusions under the HFRA to avoid NEPA review in 72 percent of all "vegetation management" projects.[19] Federal policy concerned with natural infrastructure would reverse that trend and make NEPA review a regular part of forestry decisions, including vegetation management projects under the Healthy Forest Restoration Act. Similarly, the departure from ecosystem service values, best science, and public participation in forest management plans must be corrected, perhaps by reinstating the Clinton-era rules issued in 2000. Finally, as the example of the "roadless rule" suggests, the OMB's methods for evaluating regulatory efficiency must change to better reflect the considerable nonmarket benefits of preserving natural services like flood control and biodiversity. When the only "cost" of demolishing a forest is the price of the logging road, we're in trouble.

WATER MANAGEMENT

Forest management is important to disaster control. But our nation's real embarrassment involves the management of navigable waters. Government water projects, from reservoirs to dams to levees, have radically altered our rivers and coastal waterways in ways that destroy nature's capacity to control floods, mud slides, and storms. These projects are often intended to encourage residential or commercial development. Environmental consequences, including the degradation of protective infrastructure, are ignored or discounted in the process. The Army Corps of Engineers is responsible for the lion's share of these initiatives, including the construction and maintenance of more than 140 ports, the construction of 11,000 miles of inland navigational channels, 8,500 miles of levees and floodwalls, and more than 500 dams.[20] The Corps also manages shoreline protection and restoration, construction of seawalls and jetties, and the rebuilding of beaches.

As these numbers suggest, the Corps's mission has expanded over the years. First created as an agency of military construction during the Revolutionary War, by the early 1800s the Corps was put in charge of taming America's rivers to facilitate commercial transportation. As the agency's web site explains:

> America was a young nation, and rivers were its paths of commerce. They provided routes from western farms to eastern markets and for settlers seeking new homes beyond the Appalachian frontier. The rivers beckoned and enticed, but then could treach-

erously destroy the dreams of unwary travelers and shippers whose boats were punctured by snags and sawyers or stranded by sand-bars. Both commercial development and national defense, as shown during the War of 1812, required more reliable transportation arteries. Out of those unruly streams, engineers carved navigation passages and harbors for a growing nation.[21]

Travelers and shippers, of course, were not the only ones vulnerable to a river's treachery. In the Great Mississippi Flood of 1927, the swollen Mississippi River leapt its banks and split levees in 145 locations. Twenty-seven thousand square miles were submerged, 246 people killed, and an estimated 700,000 people displaced.[22] After that much-publicized catastrophe, flood control became a national concern. In the following year, the passage of the Flood Control Act of 1928 announced the federal government's assumption of responsibility for providing flood control along the lower Mississippi River.[23] The Flood Control Act of 1936, which followed a disastrous spring flood in the Northeast, explicitly made flood control a nationwide responsibility. The act gave the Corps broad discretion to construct projects anywhere in the country "for the benefit of navigation and the control of destructive flood waters and other purposes," with only three limitations: the benefits had to outweigh the costs; host states had to contribute the land; and Congress had to appropriate the funds.[24] The good news was that Congress had formalized a national role for flood control. Floods, after all, have caused more damage to the country than any other type of natural hazard; and their effects ripple across state borders and the national economy. But the bad news was that the Flood Control Act also led to corruption and waste by opening the door to many pet projects that were not necessary for, or even helpful to, the goal of flood control.

The prominent New Deal historian William Leuchtenburg has put it less graciously, calling the act "ill conceived and wretchedly drafted."[25] Tulane University law professor Oliver Houck notes that the act's ambiguous provisions "opened a huge candy store" for special interests hoping to benefit from government works.[26] Its mission to provide "flood control and related purposes," he writes, was soon "swallowed by the 'related purposes.'"[27] By the 1960s, the "country was being dammed, drained, pumped, and leveed, by hundreds of Corps projects feeding real estate development, energy production, soy bean crops, and right on down to recreational lakes."[28] In this feeding frenzy, flood control proper became just one of many missions—a "bridesmaid" (never a bride) in Houck's view.[29]

What about the notion of using nature's own defensive structures to minimize risk? As the Corps historian Joseph Arnold writes, the 1936 Act "largely ignored . . . nonstructural alternatives," instead favoring "vast construction projects that have in a number of cases been questioned by water resources experts."[30] In fact, the Corps has been criticized on many fronts for building expensive and environmentally harmful projects. Reports from the National Academy of Sciences, the Government Accountability Office (GAO), the U.S. Inspector General, the U.S. Commission on Ocean Policy, and other groups with similar expertise have chronicled what Senator Russ Feingold has called a "pattern of stunning flaws" in Corps projects over the years.[31] These include routine overestimation of economic benefits and underestimation of environmental harm, a process strongly biased toward project approval, and a failure to follow through with environmental mitigation.[32]

In 2006, the GAO examined the Corps studies of four major water projects from Oregon to Delaware. In a damning report, the agency found that these project studies were "fraught with errors, mistakes, [and] miscalculations, and used invalid assumptions and outdated data." Many of the studies' assessments on which many lives and millions of dollars depended "did not provide a reasonable basis for decision making."[33] What is more, the agency's system of internal review seemed incapable of catching such basic, yet important, errors. These problems, the GAO concluded, "clearly indicate that the Corps' planning and project management processes cannot ensure that national priorities are appropriately established across the hundreds of civil works projects that are competing for scarce federal resources."[34]

While these projects involve mistakes in design and construction of artificial infrastructure, they also raise questions about the assessment of the foundational natural systems, whether water flow, natural floodplains, or climate. In its love affair with concrete, the Corps has spurned the steadfast strength of Mother Nature. The aforementioned inspector general report on the Upper Mississippi specifically criticized the agency's institutional bias in favor of large-scale structural projects rather than more passive natural systems. After the Midwest flood of 1993, which left more than 75 towns and 10,000 homes under water, an interagency review faulted the Corps for relying too much on earthen levees and not enough on the overflow storage services of the natural flood plains (which at the Corps's invitation had been opened for development). The report urged Congress to require the Corps to give full consideration to "nonstructural flood damage reduction alternatives." The Corps has also ignored the more threatening side of ecological systems like climate change. A report of independent engineers, for instance, worries that the agency's Comprehensive Everglades Restoration Plan lacks the surface area to hold the increased amount of rainfall that many expect from global warming. The authors fear that resulting overflows would flood estuaries on the Caloosahatchee and St. Lucie Rivers, causing them even further damage.[35]

The Corps, of course, is not the only government agency that has undermined nature's impressive coastal infrastructure. As part of the Central Valley Project, the Bureau of Reclamation helped drain the vast wetlands of California's Central Valley and channelized its rivers. The project destroyed 95 percent of the Sacramento River delta, which is the hub of the state's water supply system and hosts infrastructure critical to the Bay Area's economy, as well as being home to half a million people and a large agriculture industry. Much of the land in the region has subsided below sea level and is protected only by an aging system of levees built by the Army Corps. River floods, earthquakes, and climate change all threaten this fragile system. According to California's Department of Water Resources, a strong earthquake (magnitude 6.5) in the delta region could topple levees and flood the city of Sacramento with 17 feet of water. Salt water from San Francisco Bay would stream into the lowlands, disrupting state water supplies, cutting off transportation, and halting agricultural operations for months. Transportation and agriculture. Thousands of jobs would be lost and tens of billions of dollars.[36] Months after Katrina, a nationwide levee review conducted by the Corps found that 36 levees in California's "Sacramento District" have an unacceptable risk of failing.[37] (Nationwide, the review found that 122 levees in 30 states fit this category.) In a region notoriously prone to seismic activity, designers of the Sacramento system never even intended it to survive a major earthquake.

One of the most dramatic attempts to improve storm protection by improving natural infrastructure is now happening in Louisiana. The project is far from perfect, but it shows the potential for this new approach and can help us identify future opportunities for improvement. Thus, having flown across Sacramento, the Midwest, and the Caloosahatchee, it is time to dip the plane down over the Atchafalaya and Mississippi River deltas and see what Louisiana is doing to build a "sustainable coast."

We have seen how important coastal features are in protecting communities all over the world from storms. Viosca wrote of the importance of wetlands to storm protection in the Louisiana bayous. Native Americans appear to have understood the process a thousand years earlier. Yet in most of its history of flood control, before 2005, the Army Corps had never given natural infrastructure much thought. It had always relied on levees, locks, and floodgates. While coastal erosion had for decades been a source of concern to local scientists, engineers and the Corps had seldom focused on the issue. A Corps geologist named John Lopez changed all that. In 2002 Tropical Storm Isidore nearly flooded his house on Lake Pontchartrain, and the experience led him to wonder if the region's sinking marshes had played a role. As journalist John McQuaid tells it, Lopez began researching the issue on his own time, "examining every feature of the delta landscape that could possibly offer protection from a surge wave, starting with the continental shelf and ending with evacuation routes."[38] He looked at barrier islands, marshlands, elevated highways, even fish camps on stilts, and considered their role in dampening surge. By the spring of 2005, Lopez had crafted an idea he called "multiple lines of defense," which McQuaid describes as the "radical yet intuitively simple new approach built on the notion that all elements involved in flood control must work together—or they won't work at all."[39] That spring Lopez pitched his thesis to the chief engineer of the Corps's New Orleans district, but was politely brushed off. Two months later Katrina swamped the city.

Several years later, Lopez's "multiple lines of defense" strategy lies at the center of a new coastal protection plan for the state, which will serve as the framework for the Corps's own proposal for coastal protection and restoration. The plan, called the Comprehensive Master Plan for a Sustainable Coast, was developed by a special interdisciplinary team and shuttled through dozens of "stakeholder, science and engineering, and public meetings."[40] It is billed as "the first plan in Louisiana's history designed to fully integrate the protection of coastal assets and infrastructure with the restoration of the state's rapidly eroding coastal wetlands."[41] But that is too modest. In terms of cost, complexity, and geographic scale, this could be the largest "integrated" storm-protection plan in the *world*.

First, the document of the Comprehensive Plan reviews the need. It recounts the alarming disappearance of coastal marsh and notes that Louisiana has lost more than 1.2 million acres of coastal land, at the rate of 15,300 acres per year, since the 1930s. "Pipelines, navigation channels, and fisheries as well as centuries-old human settlements and priceless ecosystems" now hang in the balance.[42] It reminds readers of how Hurricanes Katrina and Rita deepened the gouge, how 200 square miles of wetlands were demolished, how the storms disrupted the national economy, caused a spike in fuel prices, and crippled grain shipments to world markets. What is more, those storms destroyed 200,000 homes, dislocated a million people, and killed more than 1,400 Louisiana residents. Hurricane protection in the Gulf, the authors write, must become a state and national priority. "We are living in a historic moment, one that presents us with a stark choice: either make the bold and difficult decisions that will preserve our state's future, or cling to the status quo and allow coastal Louisiana and its communities to wash away before our eyes."[43]

Storm protection was the genesis of the plan, but it is also concerned with other goods and services nature provides. By reclaiming the coast, the authors hope to improve water filtration as the Mississippi's nutrient-filled waters slide into the Gulf. They hope to preserve vital habitat for recreation and commerce, emphasizing that "Louisiana provides 26 percent (by weight) of the commercial fish landings in the lower 48 states" and that "[m]ore than five million migratory waterfowl" winter in Louisiana marshes.[44] Reminding us of Thoreau's swamp-induced raptures, the plan also highlights the cultural and contemplative side of Louisiana's soggy savannahs, promising to protect, "to the extent practicable," the state's "historic properties and traditional living cultures and their ties and relationships to the natural environment."[45]

The plan purports to "emphasize sustainability of ecosystems," to "integrate flood control projects and coastal restoration initiatives," and to create a "living document" that can adapt as landscapes change and technology and knowledge advances.[46] The authors understand that a sustainable landscape is the backbone of all other goals and explicitly subscribe to John Lopez's "multiple lines of defense" strategy. But in language resembling the vagueness of "multiple use," the plan declines to "elevate one set of needs over another" and promises instead to balance all "cultural, economic, and ecological" interests.[47]

The plan calls for dozens of projects that would span hundreds of miles and take nearly a generation to complete, restoring important elements of the Mississippi and Atchafalaya deltas and embedding reliable hurricane protection along the coast. On the Mississippi River, a set of giant diversions would be built to channel the river's sediments and freshwater to nourish existing wetlands and build new "delta lobes." In addition, millions of tons of sand from either the river bottom or the Gulf would be pumped through miles of pipes to restore marshland important to habitat, storm protection, or both. Barrier islands like the Chandeleurs would be similarly bulked up, and some turned into preserved wildlife areas. Important shorelines would be stabilized with trees, rock structures, and living reefs. "Mr. Go" (Mississippi River–Gulf Outlet: MR-GO), the commercial channel that many blame for funneling surge into New Orleans, would be gone. To the west, on the Atchafalaya River delta and Chenier Plain, the Comprehensive Plan gives many of the same prescriptions, emphasizing water management programs to keep marshes flush with freshwater to keep lethal saltwaters at bay.

In protecting against hurricane, the Comprehensive plan calls for a new and upgraded system that integrates natural and artificial features, with "[t]he level of protection provided" to be "proportional to the assets at risk."[48] This is a recognition of the idea that we must live within the reasonable bounds of infrastructure, natural and artificial. We can't expect infinite protection. And we should not taunt fate by moving valuable assets to dangerous places on a whim. But the fact is that these assets now exist where they are. They were put there not on a whim but because the services of navigable water, fossil fuels, and plentiful fisheries promised to support large sectors of the American economy, benefiting citizens from Seattle to Sarasota. In addition, many such locations were not as exposed to natural hazards as they are today because—partly as a result of public and private development—the land has subsided and the natural storm barriers have eroded. Where engineering and science suggest we can defend these areas at a reasonable economic and social cost, we will. (I'll address what's "reasonable" in the discussion of risk management in Part III.) But the plan also prepares residents for the reality that in addition to natural and artificial features, other nonstructural tools like insurance, zoning, elevation requirements, building codes, and better evacuation routes must also be folded into the mix.

Among the many levee projects on the table are a barrier system to protect the north shore of Lake Pontchartrain, a federalized levee system in Plaquemines Parish (which contains the state's bird's foot peninsula), "ring levees" around communities near Lake Charles, and a controversial proposal to construct a previously approved line of ramparts in Terrebonne Parish and the Atchafalaya delta known as "Morganza to the Gulf."[49] The document of the plan offers no price tag. But an educated guess puts it somewhere between $14 and $40 billion over the next 30 years.

On its face, there is a lot to like for advocates of natural infrastructure. The proposal endorses a *holistic* approach to flood control, integrating artificial and natural components into a multilayered plan of defense. Redundancy, a hallmark of many natural systems, is retained and expanded. The system is also *adaptive*. The plan relies heavily on an adaptive "science and performance based process" that allows designs, construction standards, and even policy priorities to evolve over time as scientific knowledge and geographic conditions change. It also expands a coastal monitoring system to ensure that changes in soil composition, water salinity, and habitat use can be incorporated into existing plans. This is important because sometimes flood control projects produce unintended consequences in surrounding ecosystems. The Dutch learned this lesson the hard way when, a generation ago, scientists discovered that offshore barriers were killing fish nurseries by blocking tidal flow. Engineers are still battling to restore vital ecosystems on Holland's western coast. Adaptive strategies also help minimize the risk of hard-to-predict external events like changes in climate, storm behavior, and relative sea level.[50] On this point, the Comprehensive Plan again reminds us of the Dutch, whose system of adaptive climate and storm management is one of the most sophisticated in the world. Advocates of natural infrastructure should also approve of the plan's expanded role for public participation and its stated commitment to "living traditional cultures." These aspects recall the values of positive liberty and distributional fairness that I drew from Joseph Singer in Chapter 3.

Soon after a draft of the Comprehensive Plan was released, it was met with an anvil chorus of criticism. Environmentalists wanted more wetlands. Residents wanted more levees. Nearly everyone, it seems, wanted a stronger standard of protection than Congress had originally authorized. That standard of protection was intended to withstand a 100-year storm, or more accurately, a storm with a 1 percent chance of occurring in any given year.[51] Katrina, by comparison, was roughly considered a 300-year storm, that is, a storm with a 0.33 percent chance of occurring in any given year. State officials revised the plan and redrew the maps. Some critics were satisfied, but most were not. In an unusual response, a team of engineers and scientists, including Lopez himself (the Multiple Lines Defense Team) assembled an entirely different plan that they say more closely follows the "multiple lines of defense" strategy. Environmentalists and natural scientists tend to prefer the "Lopez Plan"; residents in areas left unprotected by the Lopez Plan hate it. The rest of the public appears either ambivalent or unaware of the Lopez Plan. They just want a plan selected and implemented—quickly. Ultimately, the choice of protection and restoration projects will be made through a process that is still being developed and is overseen by a new federal interagency task force led by the White House Council on Environmental Quality.

The Multiple Lines Defense Team's alternative plan would better integrate the separate coastal projects, including levees, floodgates, river diversions, and marsh creation. This plan would rely more heavily than the Comprehensive Plan on the ecosystem's health as a factor in evaluating individual restoration and barrier projects.[52] The alternative plan proposes a much more aggressive use of water diversions and spillways to nourish wetlands and build new landmasses. That would mean drawing significantly more water from the Atchafalaya and Mississippi rivers; the Team argues that this can be done without disrupting commercial shipping. Their plan would also emphasize new building codes and comprehensive planning to ensure that development is safe from storm surge and does no further damage to fragile wetlands. As it stands today, local governments are not required to engage in planning and zoning, and many lack the resources. The Multiple Lines Defense Team argues that levees should be limited to assure that they do not cut off the flow of water and wildlife to wetlands. In addition, levees should not encourage humans' development of wetlands by including them in the protective zone.

One of the most important objections to the Comprehensive Plan concerns the "Morganza to the Gulf" project in Terrebonne Parish and the Atchafalaya delta. This project, proposed before Katrina, seeks to provide protection to the city of Houma and the surrounding area by building a massive levee to block storm surges originating from the south. The state and the Multiple Lines Defense Team agree that a levee should be built. The question is where to put it. To visualize the problem, imagine a horseshoe lying flat on the sand. Its open end is facing the Gulf (south), and its luck running out. Pretend the horseshoe is a natural ridge surrounded by low-lying marsh and that its contours have been sculpted by millennia of floods and shifting tides. Because the ridge is elevated, most people (though not all) live and work on the surface of that horseshoe. For the most part, the natural elevation serves them well and protects them from storms and dangerous surges: when water flows toward them from the Gulf to their south, it is usually captured and absorbed by the wetlands in the horseshoe's interior. But sometimes the surges are too powerful and threaten to sweep over the horseshoe itself. Residents, therefore, want an artificial barrier.

Where should it go? You might imagine two options. One option would be to build a levee to follow the curved half of the horseshoe (along the southern rim of the curve). Building the levee on the ridge itself gives you maximum height. When surges push inward from the Gulf, most of the time they will be naturally absorbed by the wetlands, as in the past. But when a particularly strong surge enters the marsh, the levee walls will keep water from spilling over and flooding the population centers. It is not a perfect solution: following the ridge's natural curve might increase the levee's length, which raises costs. In addition, you are limiting the area of protection to the surface of the horseshoe (and the area to its north), so you are denying protection to a minority of lowland residents, and you are limiting options for more southward development (toward the horseshoe's ends).

The second option would attempt to avoid these disadvantages by building a levee that ignores ridgeline contours and blocks the horseshoe's entrance farther south (toward its ends). This option opens up for development all the land to the north of the levee, including the horseshoe's marshy interior. If you were concerned mostly about minimizing the levee's length, you would draw the line straight from west to east. If you wanted instead to maximize land development, you could curve your line outward and then inward, closing the horseshoe with a U-shaped barrier. The levee would have to be higher, and perhaps stronger, than the first option because it would not be on an elevated ridge and would not benefit from the absorbent wetlands (which would now be within the levee's protective zone). In addition, this levee might require fancier engineering, since it would be built on looser soils than those naturally provided by the ridge on the horseshoe's curve. One thing more: the marshland, now within this levee's boundaries, would have a far smaller chance of survival, because it would be cut off from its natural source of sediments and nutrients. Because of that, it would eventually start to sink. But you might accept all that in exchange for the prospect of having a fantastic amount of new development space. If the levees were designed with special culverts and gated diversions (so-called leaky levees), you might even be to save some of the interior marshland by hitting it with regular doses of coastal water.

The Atchafalaya's geography is, of course, more complicated than this. Most obviously, the ridgeline in question looks more like the McDonald's "golden arches" than a single horseshoe. The ridge's surface area is broader, too. (The city of Houma, for those still keeping track, lies north of the arches.) But the physics are the same. The designers of the Comprehensive Plan— which is supported by many local residents—want to block the entrance of the Atchafalaya Basin with a meandering W-shaped barrier that hugs the southern shoreline. Supporters of this option are attracted by the prospect of having more dry land for development. They also want to protect the small (and somewhat charming) fishing villages to the south, which are located along Bayou Grand Caillou and Bayou Dularge. Supporters would insert a series of gates and culverts into the levees to allow access to the interior for fish and other wildlife.

The Multiple Lines Defense Team is skeptical of a hurricane protection plan based mainly on artificial barriers, especially as climatologists warn of ever-swelling seas. Lopez's team calculates that "Morganza to the Gulf" would impound more than 159,000 acres of coastal wetlands, depriving them of natural water flow and leaving almost no buffer between the levee and the Gulf of Mexico.[53] They dismiss the "leaky levee" strategy as too speculative and accuse the state of abandoning the multiple lines strategy. The "Morganza to the Gulf" project, they write, "is truly a project molded from an era of incomplete integration and inadequate engineering, permitted by the outdated notions that levees alone are adequate and that traditional levee designs will not fail. The tax payers who might pay for this project and the residents who might choose to live within this levee alignment deserve more than an outdated and ineffective flood protection system design."[54] Lopez concedes that the Team's levees would leave many villages along Bayou Grand Caillou and Bayou Dularge exposed. But their plan would protect the more populated stretches of those water bodies with more modest and less disruptive ring levees.

The Multiple Lines Defense Team has it right. While both plans subscribe to many of the principles of the natural infrastructure approach, the Team's plan does more to ensure the protection of the natural services that protect from storm and feed the fishing industry. Where there are reasonable choices to be made, policy-makers should err on the side of the systems we know work best: nonstructural systems like planning and zoning and systems based on natural structures that provide multiple benefits to multiple constituents.

Left out of the discussion so far is the subject of money. I said earlier that the cost of Louisiana's coastal restoration was uncertain, but that would be at least $14 billion over several years. That is real money, to be sure, but not mind-boggling. Boston's Central Artery/Tunnel Project (the "Big Dig") had cost more than $14 billion by 2006. What a Harvard president once said about education is equally true of infrastructure: if you think it's expensive, try living without it.[55]

The real issue for a country as wealthy as the United States is not finding the money but making sure that it will continue to be available once the project has begun. So far, funding for this long series of projects has been stitched together from a variety of sources—$50 million a year through the Coastal Wetlands Planning, Protection, and Restoration Act, a few hundred million through the Energy Policy Act, nearly $6 billion through an earmark in a war spending bill (conditioned on the state contributing $1.8 billion in matching funds), and so on.[56] But such a patchwork discourages consistent planning. Each year, one reads reports of life-saving storm protection projects being put on hold because federal funds or state matching funds are unavailable.[57] What is needed is a large and reliable funding source dedicated to natural infrastructure projects like coastal restoration.

Louisiana's coastal restoration effort, along with the other examples of flood control mentioned earlier, suggest opportunities for nationwide reform. First, the flood control mission should be promoted from bridesmaid to both bride and groom. One way to put flood control on the top of the wedding cake would be to create a separate agency whose *primary* mission would be to survey flood control needs and propose appropriate strategies. The agency would work closely with the Army Corps, the EPA, and other agencies in developing policy. Short of this reform, Congress should revise flood control policy by charging the Army Corps, the Federal Emergency Management Agency (FEMA), and other agencies to develop a *nationwide* and *long-term* (say 50 years) strategy for minimizing the risks of storms and floods. Just as important, Congress should create a dependable funding mechanism for providing support for the most urgently needed projects.

Second, the agency or agencies charged with flood control policy should be directed to develop strategies that emphasize natural structural controls (like wetlands and barrier islands) and nonstructural controls (like zoning changes and flood easements). Also, Congress should substantially reform the Corps's processes. New legislation should ensure that its projects are environmentally and economically sound and reflect the goals of long-term sustainability. The Corps should develop and follow uniform standards for shoreline restoration projects that survey the full range of options and that prefer systems that integrate natural services over those that do not. In short, the Corps should become a vital force for environmental restoration that works easily and enthusiastically with other resource management agencies. There are signs that the Obama administration may be moving in the right direction. In 2009, for instance, the White House Council on Enivronmental Quality released a proposed revision to the federal guidelines that govern water resource projects like navigation channels and levees. The proposed guidelines would for the first time require responsible agencies to "give full and equal treatment to nonstructural approaches" and consider environmental protection as a goal "co-equal" with economic development.[58]

Public Regulation

Regulation differs from public stewardship in two ways. First, regulation applies not only to government actors like the Army Corps or the National Highway Traffic Safety Administration, but to private actors as well. Second, as a result of the first difference, regulations involve the exercise of legislative authority rather than authority derived from public ownership. Over the last half century, environmental regulation has brought dramatic advantages to current and future generations of human beings and to the rest of the planet as well. But, despite its success, the regulatory regime responsible for pollution control and natural resource protection has not kept up with scientific insights about natural infrastructure.

Let's look at wetlands protection and coastal zoning, two subjects of vital interest to anyone interested in naturally triggered disasters. On the federal level, wetlands are protected under a program established in section 404 of the Clean Water Act. The program requires (with some exceptions) a permit from the Army Corps of Engineers for the discharge of dredged or fill material into "waters of the United States," including wetlands. The program requires developers to "mitigate" wetlands damage as a condition of receiving a permit. Mitigation requires developers to make efforts to avoid damage, minimize damage, or compensate for damage by building or restoring wetlands elsewhere; the efforts must be made in that order but need not go beyond what is "appropriate" or "practicable." In considering mitigation requirements, the Corps is encouraged to consider many functional values of wetlands and has clear authority to factor in services like water storage and flood control. In fact, in cases of compensatory mitigation, the Corps is required to consider the "functional values" of wetlands to be restored or replaced and to favor a remedy on the same site.

All this looks good from an infrastructure perspective, but things get lost in the details. While agency guidelines mention several times the relevance of "functions" and "values," these terms are never defined and are never given a particular weight or priority in the mitigation calculus. "These open-ended provisions," write Ruhl, Kraft, and Lant, "have led some observers to describe [the mitigation guidelines] as providing the Corps 'virtually unfettered discretion in determining whether a just compensation for destroyed wetlands has been achieved.'"[59] The situation is no better regarding state laws. Many states have statutes and administrative protections for wetlands, and many of those laws explicitly require wetlands functions to be considered in decision-making. But most state programs do not impose minimum standards for wetlands functions or otherwise "use ecosystem service values as explicit decision-making criteria."[60]

To see the danger of ignoring wetlands, one need only think back to the Iowa floods of 2008, in which 36,000 people lost their homes and 83 of 99 counties were declared disaster areas.[61] Soon afterward, scientists began suggesting that much of the damage might be attributable to "a landscape radically engineered by humans," where absorbent prairies have been plowed, creeks straightened, and flood plains developed.[62] As is the case in many states in the Bread Basket, 90 percent of Iowa's wetlands have been destroyed.[63]

As we have seen, protection against ocean storms and tsunamis greatly depends on a shoreline's natural infrastructure, including stable sands, coastal forests, and healthy marsh. The same elements also provide another category of services that, for better or worse, have a much higher value in today's marketplace: scenery and recreation. All those sand dunes and swaying palms bring lots and lots of people, along with their condos, golf carts, and Home Depots, and pretty soon the very things they liked most about the coast (the spacious beach, the lush estuary) start to get crowded and degraded. But another essential set of infrastructure that they may not even think of (the tangled beach vines, the pools of stinky peat) are also disappearing. Land use in the United States is traditionally a local affair, but in 1972 Congress established a federal role for land management on the coast, a testament in part to the public's special affection for sand and surf. The Coastal Zone Management Act encourages states to conserve the nation's coastal resources, including wetlands, floodplains, sand dunes, barrier islands, and coral reefs, as well as the fish and wildlife living in those areas.[64] The law offers federal money to any coastal state or tribe (including states and tribes in the Great Lakes region) to develop and implement a "coastal management program," so long as that program adheres to a set of federally established conservation criteria. Once approved and in place, the coastal management program governs the development activities of not only state and private actors but of the federal government as well.

In their evaluation, Ruhl, Kraft, and Lant allow that such local manage-

ment programs "are generally regarded as being effective in making contributions to the national objectives."[65] But these authors find only a vague
commitment to the functional services that coastal structures provide. The
federal criteria, for instance, require coastal management programs to balance conservation and development interests in the coastal zone and to
include procedures for preserving or restoring important features "for their
conservation, recreational, ecological, historical, or esthetic values," but dis-
appointingly provide no standards for doing this.[66] Thus these authors are
not surprised to find that while state coastal management programs will
occasionally pay lip service to ecosystem functions in their introductory
statements, "a substantial majority" of them decline to go further.[67] As an
example, the authors cite New Jersey's coastal protection statute, which
celebrates the estuary's role in "protect[ing] the land from the force of the
sea" but fails to make this function an explicit consideration in deciding
whether to develop such an area.[68] In contrast, two states, Florida and Massachusetts, offer examples of what a program more concerned with protective infrastructure might look like. Ruhl, Kraft, and Lant point out that in
Florida, anyone wishing to build in areas near the shore must show that the
proposed development will not impede "the existing ability of the [beach
and dune] system to resist erosion during a storm."[69] Similarly, Massachusetts will not permit federal development in its coastal zones without a
promise to protect "the beneficial functions of storm damage protection
and flood control provided by natural coastal forms."[70]

These are just some examples of how environmental regulation fails to
protect the natural infrastructure that is important to disaster management.
Like the law of public management, the law of public regulation suffers
from a lack of detailed priorities at the implementation stage. Legislation
and agency rules should be revised to emphasize the ecosystem function in
wetlands and shoreline protection. As others have argued, ecosystem functions should also be more deeply integrated into the NEPA process, the
protection of endangered species, and the regulation of public and private
forests.[71]

Economic Incentives

Economic incentives can be divided into subsidies and taxes. Subsidies encourage people to mend their ways and adopt desirable behavior, paying
them, as some have put it, to "go straight." Taxes discourage people from
acting in undesirable ways by charging them instead—that is, "scaring them
straight" (or at least making sure that they pay their dues). Some environ-

mentalists are skeptical of subsidies because they dilute the moral imperative of good stewardship and can perversely lead actors to threaten *bad* activity unless they are paid to do good. What is more, government subsidies come with a very checkered past. In the last generation, agribusiness, the timber industry, and oil companies (to name just a few) have wangled tens of billions of dollars from Congress to help drain wetlands, mow forests, and punch holes in Alaskan tundra.[72] To take just one example, the United States indirectly subsidizes logging in national forests by selling timber at below-market rates and by providing logging roads at public expense. For these and other reasons, the U.S. Forest Service operates at an annual loss of nearly $1 billion.[73] While reliable numbers are hard to come by, the General Accounting Office has estimated that between 1992 and 1997, the Forest Service lost $2 billion in below-market sales alone.[74]

Still, green subsidies have a place in protecting natural infrastructure. They seem particularly useful where an industry's political influence is so substantial that it forecloses direct regulatory options. (In this way green subsidies address both the problem of market failure *and* government failure.) Advocates of ecosystem services are particularly enthusiastic about trends in American farm subsidies that encourage farmers to take high-value natural infrastructure like wetlands and floodplains out of production.[75] The Millennium Ecosystem Assessment champions green subsidies in both agricultural and nonagricultural settings as a way of avoiding contentious fights about regulation in many countries.[76] In the United States, subsidies could be used to supplement existing regulations that prove underprotective. The government could pay landowners to conserve urban wetlands or to locate mitigation projects closer to a depleted resource.[77]

Subsidies also show promise where there is a small, well-heeled group of beneficiaries who could be motivated to pay for resource recovery if given a mechanism for doing so. Thus John Forgrach, an insurance executive and entrepreneur, made news when he proposed selling 25-year bonds to finance a reforestation effort in the Panama Canal region. The canal relies on an enormous flow of freshwater (about 40 million gallons per ship) to operate. But over the years this water has been drying up because of deforestation and rising temperatures. This slows the course of cargo ships through the canal. Forgrach hopes to recruit shipping clients like Wal-Mart and auto manufactures, who currently insure against damage from a canal closure, to invest in reforestation in exchange for lower insurance premiums.[78]

Taxes are the flip side of subsidies. In theory, such instruments can help conserve natural infrastructure by "provid[ing] negative incentives that will sustain optimal levels of [resource use] in a competitive system."[79] In contrast to subsidies, which make "beneficiaries" (the citizenry, Wal-Mart) pay, environmental taxes are grounded on the principle that the "polluter" should pay. The challenge in designing this kind of tax policy is that to do it right, one needs reliable science. If scientists underestimate the degree of environmental harm an activity causes (including the cost of restoring it), policy-makers will "undercharge" for the activity; if scientists overestimate the environmental harm, policy-makers will overcharge. Scientific uncertainty causes political problems, too. A polluting industry might disagree with what scientists say its share should be, or it might disagree with the choice (and expense) of remediation efforts. Politicians, depending on their persuasion, will greet these complaints with either sympathy, skepticism, or a very unscientific desire to split the difference and go home.

There is also the matter of where the tax revenue goes. If the revenue goes into the government's general funds, green taxes serve the purpose of discouraging environmentally harmful activity but do not guarantee that the revenue will be used to repair the environmental harm that has already occurred. Even achieving the first purpose—discouraging environmentally harmful activity—can be tricky. Set the tax too high, and industry overregulates itself, refraining even from types of development that do not seriously threaten environmental interests. Set the tax too low, and industry underregulates itself, choosing harmful levels of development. Green taxes are less common in the United States than in Europe; but there are examples of American laws that tax industry in order to pay for the environmental damage it creates. For instance, the Surface Mining Control and Restoration Act features a reclamation fund designed to pay for the cleanup of abandoned coal mines.[80] The fund is financed by a tax on coal. The Superfund law, which facilitates the cleanup of abandoned toxic waste sites, relied for most of its existence on a tax on oil and chemicals to finance the cleanup of polluted sites whose responsible parties were either insolvent or unknown.[81] After the *Exxon Valdez* disaster, Congress created a $1 billion trust fund to supplement cleanup efforts undertaken by companies that have caused oil spills in U.S. waters. The fund, established under the Oil Pollution Prevention, Response, Liability, and Compensation Act, is financed by a tax on oil of 5 cents per barrel. In each of these statutes, the tax scheme is only one tool in the toolbox; they also depend heavily on regulation and liability rules for the accomplishment of their objectives. In addition, the taxes are assessed according to very simple formulas; there is no effort to tie the amounts to detailed scientific data. Nonetheless, generally speaking, these tax regimes worked and accomplished much of what they were designed to do. Could a tax strategy like this effectively support a more complex and sustained effort at resource restoration? I think so. To see how, let's return to the Louisiana coast.

For decades, Louisiana's constitution has dedicated revenues from state mineral leases to a "trust fund" for wetlands conservation and restoration programs. This has the effect of making sure that the monies earned by the state from environmentally destructive activity are used to mitigate that harm, a kind of "toll," if you will, on the use of green infrastructure. It is an elegant idea, intended to address imperfections in both actual and political markets. The scheme addresses the market failure problem by explicitly acknowledging the hidden, noneconomic costs of wetlands loss and linking them to extraction-based revenue, which would otherwise be used for other things. It addresses the government failure problem by ensuring a designated stream of conservation revenue that is immune from the whims of politicians and the uncertainty of annual budget battles. But while the revenue helps, no one believes it comes anywhere near compensating the state for the real value of the damage caused. Still, Louisiana deserves credit, since this compensation scheme is more than most states enact. In many mountain states, for instance, state mineral revenue is plowed into all manner of things, from schools to roads, while leaving behind poisoned and eroded mountainsides, increasing, as we've seen, the chance of dangerous floods and mud slides.

Since Katrina, the strategy of revenue capture has been pushed even further. Months after the flood, Louisiana citizens amended that constitutional provision to allow state mineral revenue to be used also for "hurricane protection." The move, born of the state's new safety obsession, highlights the practical nature of the infrastructure perspective. First, the amendment explicitly names a *function* within the natural landscape that the state wants to restore. This emphasizes the idea that nature provides services and that some services (storm protection) may be more important than others (recreation). As a constitutional provision, future courts could see this as first a step away from the empty-headed "multiple use" paradigm. Second, the amendment suggests that captured revenue may also be used for artificial infrastructure used in tandem with natural infrastructure to protect against hurricane. This acknowledges the reality that protection from natural disaster, whether storms, fires, or mud slides, will depend on a fusion of natural and artificial systems working together to protect at-risk communities.

More significant than this change was the adoption of the Dominici-Landrieu Gulf of Mexico Energy Security Act of 2006, which allows four Gulf states (Texas, Louisiana, Alabama, and Mississippi) to claim 37.5 percent of federal revenues generated from new oil and gas production on the outer continental shelf (OCS).[82] By the law's terms, the revenues must be used for coastal protection, hurricane protection, or other forms of environmental mitigation. Louisiana expects to receive the bulk of these revenues, since it hosts significantly more oil and gas producing activities than the other Gulf states. Louisiana can expect about $200 million in OCS revenues between the years 2007 and 2016. After that, new areas in the northern and central Gulf will be opened for development, allowing as much as $600 million to accrue to the state's Coastal Protection and Restoration Fund each year.[83] While most of these funds are years away, the state plans to borrow against the expected revenue in order to expand conservation efforts now. The revenue is essential to coastal restoration, since without it Louisiana would probably not be able to contribute the state "matching funds" required for its investment in Gulf Coast restoration. (If you suspect a shell game, you're right: the federal government is giving Louisiana a share of federal revenue so that the state can return that revenue to the federal government.)

This was a Faustian bargain that many environmentalists outside the Gulf region opposed. The problem was that the deal was conditioned on opening up for development previously protected areas of the OCS. For years Louisiana had lobbied for a larger share of federal revenues generated from *existing* OCS production off its shores. On that measure, Louisiana had always held the short end of the stick. Louisiana, for instance, receives only 27 percent of federal royalties generated from current production that is 3–6 miles offshore, and nothing for production beyond that. Texas, in comparison, receives 100 percent of royalties generated from current production that is 3–9 miles offshore, and 27 percent of royalties from current production 9–12 miles out.[84] Florida, another Gulf state with lots of political clout, has a similarly generous deal.[85] But even compared to interior states, the arrangement seemed unfair. Mining states like Wyoming and Utah are guaranteed half of all federal revenues from hard rock mining on federal land within a state's borders to compensate them for environmental degradation, even though they are not required to spend any of that money on mitigation.[86] The arrangement was unfair, in Louisiana's view, because although federal oil leases occurred miles away from state jurisdiction, all wetlands destruction necessary to that development occurred within the state's borders, endangering, as it happens, hundreds and thousands of state residents. After Katrina, Louisiana's congressional delegation sensed a window in which it could negotiate for a bigger piece of federal oil and gas revenue. However, it could only do so in exchange for allowing *more* oil and gas development to occur offshore, and even then the increased share would accrue *only* to that new development. It is an example, as Houck has said, of government officials wanting to "have their marsh and eat it too."

A better market strategy would be to directly tax the oil and gas industry to restore the coast they helped dismantle. But this is a strategy neither federal nor state officials get excited about. Since Katrina, Louisiana's political leadership has suggested that for the most part, the federal government should bear the costs of protecting the coast. They have launched an advertising campaign called "America's Wetland" to raise awareness of the Gulf's coastal wetlands loss and to promote federal funding for coastal restoration.[87] In New Orleans, tourists and residents see symbols of the campaign everywhere—on signs, bumper stickers, and those trendy rubber bracelets. The campaign's trademark mascots, a giant blue heron and a plush bottle-nosed dolphin, march regularly in Carnival parades. Such ecological awareness is surely welcome. On the campaign's web site, viewers learn many of the compelling facts that make coastal restoration in the Gulf an immediate national concern. They learn about the importance of Louisiana's wetlands to the fish and shellfish industries, to ports and shipping, and to "the domestic energy supply."[88] What they do not learn on this web site is that a large share of the coastal damage now in need of repair was caused by the oil and gas industry. This omission is not so striking when one learns (after three well-placed mouse clicks) that the campaign's main sponsors include Shell Oil, BP, ConocoPhilips, Exxon Mobil, and the American Petroleum Institute.

In response to this effort, several local environmental groups have begun campaigning for state legislation that would require oil companies operating in the state to contribute to the state's coastal restoration efforts. Industry representatives claim that would be unfair and counterproductive. An official from the Louisiana Mid-Continent Oil and Gas Association, for instance, has suggested that the amount of damage attributable to oil and gas is "less than 10 percent."[87] Besides, others suggest, no one knew back in the

1960s and 1970s, when most of the industry's canals were dug, that those cuts would be so damaging. This defense is implausible. Studies conducted by the U.S. Department of the Interior estimate that 30–60 percent of all destruction of the state's coastal wetlands can be attributed to oil and gas development.[90] In addition, reports by the Army Corps dating back as far as 1973 describe how the oil and gas industry destroys miles of wetlands every year.[91]

At a time when oil industry profits seem to regularly hit "record levels," a cost-capturing green tax to restore the coast would seem to make enormous sense — particularly when one considers that one of the main beneficiaries of the effort would be the oil industry itself, whose miles of pipeline and industrial equipment depend on barrier islands and marshland for protection from storms. State officials are reluctant to take on the oil industry, apparently out of fear that the industry will develop its refining capacity elsewhere. Yet this seems unlikely. The industry has invested too much in the region to walk away; the oil and gas companies would lose more than they would gain.

Property Interests

The doctrines discussed here fit in only a loose category of entitlements that often, but not always, involve land use: entitlements offered in tort doctrines like nuisance and negligence and in the public trust doctrine. These entitlements have often originated in Anglo-American common law, but many are hybrids today, codified in state statute, buttressed in state constitutions, or, as in the singular case of Louisiana, rooted in the Napoleonic Code. In this category, we might also include the takings clause of the Fifth Amendment to the U.S. Constitution, which places an outer limit on the degree to which federal or state laws can regulate private land without offering owners "just compensation."[92] In the 1990s, many feared that a series of Supreme Court decisions in this area might create a doctrine unfavorable to regulatory efforts to protect important ecosystem features like sand dunes or flood plains on private land.[93] In the decade that has followed, however, the Court seems to have retreated from that trend, at least for the time being.[94] For this reason, I will not discuss the takings clause directly in this chapter.

Historically, traditional property law was a friend to environmental preservation. Conventional wisdom holds that the common law, imported from eighteenth-century Britain, lacked the necessary mechanisms to protect the wild American landscape. But that is not the full story, as a new generation of property scholarship has demonstrated. The problem, according to these scholars, was not that traditional property law was blind to the natural world. On the contrary, property law looked Nature right in the eye and then squashed it like a bug. Law professor John Sprankling has developed this argument in an important article, "The Antiwilderness Bias of American Property Law." In this comprehensive review, he examines the evolution of American property law in contexts involving wilderness, paying close attention to the points at which American legal doctrines begin to depart from the English ones on which they were based. Sprankling finds that American courts early on found ways to bend English property law in ways that prefer agrarian (and later industrial) development over passive ownership, particularly passive ownership of "empty" wilderness.[95]

Thus, adverse possession, the doctrine by which a long-term possessor can take ownership from one who has true title, was transformed from a mechanism to clear title on the English moors to a means of rewarding trespassing berry pickers and loggers who prove more industrious than a forest's true owner. Similarly, nuisance law morphed from a protection against nearly any interference from activities on a neighbor's land into a protection against only "unreasonable" interference, a requirement that often pitted the economic loss from the plaintiff's harm against the economic utility of the defendant's activity. The upshot, according to Sprankling, was that "all other things being equal, conduct was less likely to be enjoined as nuisance if it occurred in a wilderness area than in another, more developed, locality."[96]

This is bad news for defenders of natural infrastructure because wilderness—the least protected domain in American property law—is exactly where one would expect to find the most impressive producers of ecosystem services. Why is our nonstatutory law stuck more than a hundred years in the past? Unlike the case of statutory law, where we could blame government failure (that is, pork barrel politics), the primary explanation for today's antiwilderness bias appears to be habit. According to Sprankling, today's courts seem to have forgotten that the nineteenth-century doctrines favoring development over conservation were tailored to a different era when nature seemed inexhaustible. They continue to "blindly appl[y] most of the antiwilderness doctrines of the past" without realizing their cumulative effect.[97] Perhaps more important, these precedents have insinuated themselves over dozens of generations into the very fabric of community life. "Highest and best use" is not just the mantra of the property lawyer but of every landowner in every country and parish. And, 99 times out of 100, that "use" has nothing to do with preserving natural infrastructure.

The result is that common law property doctrines—so often praised for flexibility and innovation in other contexts—have simply failed to deliver when it comes to acknowledging property rights in ecosystem services. Or as Ruhl, Kraft, and Lant put it: "American property law is not simply neutral on the question of private property rights in natural capital and ecosystem services, but downright hostile, making it no wonder that neither finds much stock in the marketplace."[98] Still, new developments in economics and science offer hope that property law might soon be better deployed to protect natural infrastructure. Property law and the ecosystems services perspective are both about the same thing: optimizing the value of owned resources for human use.[99] The main reasons that property law has not historically protected the rights of private owners or the public to ecosystem services has been lack of convincing evidence establishing the value of those services and an unwillingness by courts to move beyond familiar frameworks. But where the value is clear, courts have shown some willingness to budge. Thus a few older cases have protected a landowner's access to "natural rainfall" against unreasonable interference from cloud seeding operations.[100] More recently, the high courts of Wisconsin and New Hampshire have ruled that the common law protects a landowners' right to access "sunlight as an energy source" against unreasonable interference by neighboring structures.[101]

What about using the common law to protect ecosystem services that minimize natural hazards? This application presents some added challenges. First, as I've already pointed out, precise information about nature's protective capacity is still hard to come by. Establishing the value of sunlight and rooftop solar energy production is one thing; describing the precise value of a forest or swamp as a flood barrier is another. Second, these "natural barrier" cases differ from the "natural access" cases, in that they ask the court to hold that a plaintiff landowner has a property interest in a natural service (barrier protection) that is physically located on the defendant's land. The large majority of common law nuisance cases are cases in which a plaintiff demands that the defendant stop imposing a *negative externality* like noise, smoke, or some other irritant. There are very few cases in which plaintiffs allege nuisance based on a landowner's termination of a *positive externality* like, for instance, storm protection. When I say that protecting natural barriers through the common law will require courts to move "beyond existing frameworks," this is what I mean.

There are two reasons for cautious optimism here. First, in economic terms there is little difference between imposing a negative externality or removing a positive externality. To the extent that modern property law means to maximize value and human utility, the distinction between negative and positive externalities is irrelevant. The distinction *is* relevant to pursuing fairness and honoring the reasonable expectations of the parties. But that important goal is best promoted by the law's requirement that any interference in a plaintiff's use of land be found "unreasonable" by a court. Removing a tree on your property that has shaded your neighbor's hammock is probably not an "unreasonable" interference in this sense. Tearing up your wetlands and exposing your neighbor to a higher storm surge, however, might be.

Second, the common law tradition has already acknowledged, at least in some limited ways, a landowner's rights to positive externalities produced on another's land. For instance, Anglo-American property law has long prevented a landowner from excavating her land so as to remove support provided laterally (on the side) to neighboring property and as a result compromising that land's stability. In many states, this obligation is absolute and does not even require a finding of unreasonableness. Similarly, in many jurisdictions surface owners have an absolute right to subjacent support for their land and may prevent owners of subsurface rights from impairing their interests.[102] In coastal zones, one old line of cases restricts beach owners from excavating in ways that would foreseeably expose nearby land to the ravages of wind and tide.[103]

My optimism on this topic is "cautious" because the doctrines just described have, in modern times, been displaced by statute and are therefore less familiar and less understood than they once were. Many plaintiffs seem to have given up on private law as a means of protecting their assets from catastrophe. But one case, *Barasich v. Columbia Gulf Transmission Company*, brought after the Katrina disaster, did attempt to use private law principles in just this way.[104] Although the plaintiffs lost, their effort suggests a road map for others to follow.

The plaintiffs, a class of residents from Louisiana's southern parishes (including Orleans Parish, where New Orleans is located), sued the region's major oil companies for hurricane damage caused by "wetland loss attributable to oil and gas exploration and/or production activities."[105] The plaintiffs alleged that the companies had over several decades lacerated the state's protective marshes with thousands of miles of canals used for oil and gas pipelines and resource exploration. The canals, they alleged, destroyed millions of acres of the state's southeastern marshland, depriving inland communities like New Orleans and St. Bernard Parish of "their natural protection from hurricane winds and accompanying storm surge."[106] The plaintiffs' description of wetland damage recalled my "death by a thousands cuts" narrative in Chapter 2. The canals altered wetland hydrology by sucking saltwater into marshlands. Spoil banks choked tidal flow and blocked the delivery of minerals and land-building sediments. The area's jungly vegetation that once had stabilized soils and absorbed tidal surge died and was eventually swallowed up in the open sea. In addition, the plaintiffs alleged that the oil companies had failed to protect the canals from erosion and knowingly allowed them to develop breaks and cuts, all of which had accelerated the damage. Such damage, they argued, resulted in an "enhanced" amount of hurricane damage, for which the defendants were liable. The plaintiffs' account framed the story as the loss of protective infrastructure, a public good that had been squandered by wolfish oil companies intent on making a private buck. Of course, the former wetlands at issue where indisputably *private* property. And the oil companies that carved up that land did so under valid leases for which they had paid market value.

Mark Boudreaux, a spokesperson for ExxonMobil, called the lawsuit a "sobering reminder of our litigation-crazed society" and suggested that Louisiana storm victims were just looking for someone to blame. "It's amazing," he told the press, "how you can have an unprecedented natural disaster like Katrina and try to blame it on the oil companies."[107] A lawyer for the plaintiffs, Val Exnicios, accused the defendant companies of "callous indifference" to the marshland ecology and its role in buffering storms. "We believe it's the right time to pinpoint who's essentially responsible for the devastation caused by Katrina in the first place," he said; "the major oil and gas companies, who haphazardly dredged thousands of miles of exploration and drill site canals throughout south Louisiana."[108] Plaintiffs sought relief under Louisiana's "civil law" code; but in many respects, their claims resembled those brought under the common law doctrines of nuisance and negligence, which apply in other states. Essentially, plaintiffs argued that the oil companies had, by managing the wetlands in an unreasonable way, contributed to the foreseeable damage of their property; as a result, the companies owed them compensation.

The court rejected the lawsuit for failure to state a claim on which relief could be granted. Regardless of what the facts might show, the court ruled that under Louisiana law, the parties were too far away "in time and space" and their connection too "attenuated" to establish a duty owed by the oil companies to the residents of southeastern Louisiana.[109] That sounds bad for advocates of natural services; but there is a silver lining for those inside and outside of Louisiana. When the court ruled that the oil companies were too distant, it was relying on a rule, particular to the Louisiana Code, that holds that "nuisance" claims (as lawyers outside Louisiana might loosely call them) must involve properties that are geographically close to one another.[110] (In this case, the degraded wetlands and the ravaged properties were sometimes miles apart.) That rule does not apply in the common law nuisance doctrine of other states, so long as the plaintiff can show the damage was *foreseeable* and that it was *proximately caused* by the defendant's activity.[111] (In common law sense, proximity of cause need not be geographic.)

When the court ruled that the oil companies' actions were "too attenuated," it was making a point about cause and effect. That is, the court rejected the complaint because it "did not allege that the actions of *any particular defendant* were a substantial factor in causing the injuries suffered by *any particular plaintiff.*"[112] Instead, it appeared that the plaintiffs were trying to bundle the conduct of *all* of the defendants and to hold *each* responsible for a share of the combined damage. Louisiana does not recognize this form of "group causation." But many other states do.[113] In fact, group causation is common in environmental disputes, which often involve repeated acts of pollution or dumping by several parties over time. Thus, the plaintiffs' theory in *Barasich* might succeed in another jurisdiction, so long as plaintiffs could establish foreseeability and proximate cause.

I also said that *Barasich* offered a silver lining to Louisiana plaintiffs. That hope occurred in the end of the decision, where the court went out of its way to explain how plaintiffs might succeed in a narrower claim under Louisiana's less flexible code system. "By all accounts," the court wrote, "coastal erosion is a serious problem in south Louisiana. If plaintiffs are right about the defendants' contribution to this development, perhaps a more focused, less ambitious lawsuit between parties who are proximate in time and space, with a less attenuated connection between the defendant's conduct and the plaintiff's loss, would be the way to test their theory."[114]

Louisiana lawyers are already discussing this invitation and seem poised to test the theory sometime in the future, either in the context of Hurricane Katrina or another storm event.

In addition, we are sure to see more attempts throughout the country to protect natural infrastructure through the traditional principles of property law. The dispute over Kivalina, Alaska, that I mentioned in Chapter 3 is an example of this. The village of Kivalina sits on a small barrier reef jutting off of Alaska's northwest coast. On that frozen finger of land, a population of 400 Inupiats has lived for years, surviving on mostly caribou, bearded seal, and salmon. One key to their survival is a system of natural berms that slouch along the shallow coast, forming a ridge that is made of hard permafrost. In the fall, the snow and ice accumulate on the ridge, growing into a hardened seawall of sorts, which has historically protected the village from the savage winter storms. But over the last 10 years, global warming has slowly changed things. The permanently frozen subsoil is melting, and its valuable sands are literally sliding into the ocean. In 2004 and 2005, the village lost more than 70 feet of coastline. In addition, the icy seawalls are coming later in the season, allowing heavy storm surges to spill over the village and its 150 homes and buildings. (It, of course, does not help that higher temperatures are also causing the seas to rise.)

With few resources to protect their property or relocate, the residents filed a nuisance action in 2008 against 24 oil, gas, and electric companies that they said jointly contributed to global warming and, as a result, their melting infrastructure.[115] A federal district court dismissed the lawsuit in 2009, but an appeal is considered likely.[116] This case differs from *Barasich* in that the protective infrastructure at issue in Alaska resides on the Inupiats' own land, making this a case about imposed negative externalities (greenhouse gases that cause global warming) rather than removed positive externalities (shredded wetlands). But the case shares with *Barasich* a reliance on the group causation theory as well as a raft of complexities involving physical cause and effect.[117] *Kivalina* involves claims of both private nuisance (which protects the property rights of individual owners affected) as well as public nuisance (which protects the rights of members of the public in general who are affected), suggesting that access to natural infrastructure might go beyond the rights of individual owners. Because the *Kivalina* case could influence other cases on similar theories involving climate change and the loss of natural infrastructure, developments should be watched closely.

The public trust doctrine is another legal tool that might strengthen the public's right to preserve privately owned natural infrastructure. Born of Roman law and later migrating to English common law, the public trust doctrine has inspired more mythology than Middle Earth. The doctrine helped Caesar grow his empire and was centuries later forced on King John by the Magna Carta. During America's Industrial Revolution, its logic defended a thousand acres of Chicago's shoreline from hungry rail barons. More recently, the doctrine has been deployed (unsuccessfully) to preserve acres of dinosaur fossils in Utah and (perhaps more successfully) to save California's fabled and shrinking Mono Lake.[118] Even so, this doctrine has fallen short of many environmentalists' dreams.

At its core, the public trust doctrine holds that despite the common law's strong preference for the private ownership and development of land, there are some resources that must be forever owned and controlled by the state, to be held "in trust" for the use and benefit of everyone. By tradition, those public resources include navigable and tidal waters, as well as the land beneath them. The idea is that certain geographic features, namely fisheries and navigational channels, are just too important for any individual or private organization to control. The river, the mudflat, and the oyster bed produce far more social wealth as "inputs" to other commercial endeavors than as privately owned, service producing goods. They are, in a word, infrastructure. So the Romans, who referred to their own water resources as *res communes*—things held in common—may be credited not only with mastery of artificial infrastructure but philosophical understanding of natural infrastructure, too.

Because accessible fisheries and waterways were also essential ingredients of minimal living standards, access of such infrastructure eventually became an aspect of human rights. Thus Britain's Magna Carta, signed in 1225, reaffirmed the "liberty of navigation" and the "free fisheries," and the constitutions of several American states today include resource protections rooted in the public trust.[119] (Many nations, similarly, have resource protections in their constitutions, and I'll get to that later.) In the United States, public trust theory emerged as a free standing doctrine in the Supreme Court case *Illinois Central Railroad Co. v. Illinois*.[120] The case involved an 1869 law that granted to the Illinois Central Railroad more than 1,000 acres along the shores of Lake Michigan in Chicago Harbor, stretching a mile across the central business district and extending a mile offshore. The state legislature later changed its mind, revoked the earlier grant, and sued to have the original grant declared invalid. One can imagine the railroad's lawyers tearing their hair out and crying, "A deal's a deal." But the Supreme Court did not agree. While the state held title to the lake's bed, the Court held that given the land's special "character," the state could not sell it to private owners to the detriment of public benefit. Submerged harbor land like this, which was important to navigation, commerce, and fishing, wrote the court, must be held in trust for the people of Illinois and can "never be lost" unless conveyed for the purposes of promoting the public interest.[121] Although the Court referred to state cases extolling public sovereignty and responsibility over lands beneath navigable water, it never anchored the theory in firm precedent, leaving judges and lawyers to speculate on their own about the theory's origin and watery contours. It protects the public's interest in navigation, commerce, and fishing by prohibiting states from either transferring the beds and banks of tidal and navigable waters or refusing to regulate them.[122]

The story would perhaps end there had it not been for Joseph Sax, an audacious law professor who published in 1970 a pathbreaking article that consolidated the theory's myriad precedents and argued for its dramatic expansion.[123] This was before many of our most important environmental statutes had been passed, and Sax hoped to build a common law foundation on which a national, court-imposed environmentalism could take hold. It never happened. The Supreme Court declined all invitations to expand the theory, and Congress's mighty fleet of environmental laws enacted over the next 20 years, including the Clean Water Act of 1972, appeared to make Sax's dream less relevant. The federal public trust theory has thus remained remarkably static.

The seeds of Sax's work were scattered to the wind, but many took root in state gardens where local courts and legislatures happened to be looking for ways to fill regulatory gaps. Some states extended public trust protections to upstream waters, interior lakes, or aquifers.[124] Others expanded the list of protected public uses to include things like recreation, public safety, or resource protection.[125] State public trust doctrines are almost always grounded in more than common law. Instead, they rise from a mixture of common law, state constitutional provisions, and state legislation. Illinois, for instance, now articulates its public trust principles in case law, two constitutional provisions, and 16 statutory provisions, by one scholar's count.[126] Other states, including Louisiana, are similar in this way.

As a result, public trust doctrines among the states are more diverse than much legal scholarship has recognized. In a review of 31 states east of the

Mississippi River, environmental law professor Robin Kundis Craig finds surprising "richness and complexity" in American public trust principles.[127] Alabama, for instance, sticks to the basic *Illinois Central* rule of protecting navigation, commerce, and fishing.[128] Next door, Mississippi applies the doctrine to environmental protection, preservation, and the "enhancement of aquatic, avian, and marine life."[129] Not to be outdone, South Carolina's high court holds that its public trust doctrine guarantees not only clean air and safe drinking water but also an "inalienable right" to fish and sail.[130]

It's hard to know what phrases like this really mean. Sometimes application of the doctrine contains a substantive guarantee, that is, a promise that a particular kind of development or transfer will not take place. Other times, the guarantees appear procedural. Louisiana's Constitution, for instance, declares that "the environment shall be protected, conserved, and replenished insofar as consistent with health, safety, and welfare of people."[131] The words sound substantive, but the state supreme court has read them to require only that the government *consider* the environmental effects of its actions on balance with "economic, social and other factors."[132]

But the public trust doctrine is a good fit with environmental protection, because the doctrine is at its core about public utility and infrastructure. Sax and his followers understandably wanted more: a doctrine that might recognize aesthetics, existence value, and other less tangible public benefits. But for the purposes of protecting natural infrastructure, we do not need to stretch the doctrine beyond its first principles. All we are stretching is the concept of public utility to allow it to expand with today's scientific fact. This is a conservative use of the doctrine.[133] To apply public trust principles to natural infrastructure and disaster, we need a brief inventory of how such principles affect government action. There are four ways. First, the doctrine can be invoked to block or rescind undesirable government action. This is what happened in *Illinois Central,* when the current legislature used the federal public trust doctrine to rescind a transfer made by an earlier legislature. Second, the doctrine can encourage politicians and agency officials to pursue environmental goals—or at least give these people cover when attacked by antienvironmental lobbies. It is hard to measure the prevalence of such influence, but as we'll see, it does exist. Third, the doctrine can be used to defend desirable state action against challenges to government authority. Finally, the doctrine can be used to defend desirable state action from takings challenges.

Once again, Louisiana provides a premier example of public trust litigation as it relates to storm protection. The case involved the Caernarvon Freshwater Diversion Structure in Louisiana's Breton Basin, some 15 miles downriver from New Orleans. As imposing as the Welsh castle that shares its name, Caernarvon is a canal and gate mechanism made of concrete and steel and built into the mainline Mississippi River levee. It consists of a 175-foot-long intake structure on one end, a mile-and-a-half-long canal on the other end, and in between a mammoth box culvert that looks like five subway tubes borrowed from the London Underground. The project is designed to mimic spring floods by strategically releasing millions of gallons of freshwater from the Mississippi River into the dying marshes of Breton Basin.[134] With a capacity to divert more than 8,000 cubic feet of water per second, is one of the largest freshwater diversion projects in the world. (Its sister, Louisiana's Davis Pond diversion project, is the largest.) When the Caernarvon project opened in 1991, it was widely seen as an important experiment in coastal restoration, and in many respects it worked. In its first three years, the structure corrected the saline balance of Breton Sound and delivered significant amounts of sediment (though not as much as expected) into the basin. This process nurtured new wetlands and improved oyster habitat in many publicly owned parts of the coast. In the process, though, the diversions wiped out thousands of acres of *privately* leased oyster beds in other areas.[135] In 1994 a class of oystermen farmers, holding an estimated 204 leases, brought suit against the state. The oystermen (by convention, the term also includes women) claimed that while the state owned the submerged property at issue, the value of their private leases on that property had been destroyed, or "taken," just as if the state had condemned the leases by eminent domain. Under the state constitution, they believed compensation was due. At trial the judge agreed, and, using a controversial formula for calculating damages, awarded the oystermen a staggering $1.3 billion.[136] The state court of appeals affirmed, and soon Louisiana's "oyster wars" were front-page news and before the Louisiana Supreme Court.

Supporters of coastal restoration feared this monster judgment would crush all reasonable hope of rebuilding the state's fisheries and protective wetlands. The state was not without a legal argument. Since 1989, it had inserted "hold harmless" provisions in the oyster leases immunizing the state from damage claims based on water diversions. By the time of litigation, nearly every lease contained some version of that provision. Indeed, clauses issued after 1995 specifically linked the state's water diversions to its stewardship obligations under the state's Constitution ("the environment shall be protected, conserved, and replenished"), in addition to "the public trust doctrine associated therewith."[137] For their part, the oystermen argued that the hold harmless provisions were invalid because (in their view) the state law governing oyster leases did not authorize such immunity.

In 2004, to the relief of nearly everyone except the oystermen, the Louisiana Supreme Court threw out the billion-dollar award and unanimously reversed the ruling. Contrary to earlier rulings, the court ruled that the hold harmless provisions were, in fact, properly authorized by law and as a result barred the plaintiffs' claims. (Claims based on older leases, which did *not* contain hold harmless provisions, were also dismissed, on the grounds that those plaintiffs had waited too long to file suit.) The provisions were valid, explained the court, because Louisiana's law of oyster leasing authorized the state to make stipulations "necessary and proper to develop the [oyster] industry."[138] Because the Caernarvon project provided a net benefit to oyster production by restoring public seed grounds, the court found the hold harmless provisions to be "necessary and proper" to the goal and therefore enforceable.

This was an important development but, so far, nothing for a public trust lawyer to write home about. What happened next, however, deserves our attention. Writing for the majority, Justice Jeffrey Victory wove into his logic a discussion of the public trust. His measured tone grows more urgent as he acknowledges the fragility of his state's coastal infrastructure. Then he drops this much-cited bombshell:

> We find that the implementation of the Caernarvon coastal diversion project fits precisely within the public trust doctrine. The public resource at issue is our very coastline, the loss of which is occurring at an alarming rate. The risks involved are not just environmental, but involve the health, safety, and welfare of our people. Coastal erosion removes an important barrier between large populations and ever-threatening hurricanes and storms. Left unchecked, it will result in the loss of the very land on which Louisianans reside and work, not to mention the loss of businesses that rely on the coastal region as a transportation infrastructure vital to the region's industry and commerce. The State simply cannot allow coastal erosion to continue; the redistribution of existing productive oyster beds to other areas must be tolerated under the public trust doctrine in furtherance of this goal.

A concurring opinion in the case mentions the public trust doctrine in a footnote, this time arguing that even assuming no immunity, the plaintiffs' harm cannot be characterized as a taking because by restoring freshwater to the sound, the state was "exercis[ing] its police power to avoid a public calamity."[139] The power to avoid public calamity, the opinion goes on, is further supported by the state's public trust duties as described in Justice Victory's majority opinion.[140]

I should note that the public trust doctrine was not, strictly speaking, necessary to the court's holding. Having based the state's authority on its mission to increase net oyster production, the court theoretically could have called it a day. In fact, when the decision came down, many local lawyers dismissed the public trust discussion as folderol. But that would be a mistake. The magic of the public trust doctrine is its ability to bind onto more modern rules with epoxy-like strength in order to stabilize a controversial position.[141]

In this situation, this binding effect was used in several ways. The doctrine was first introduced by the state's Department of Fish and Wildlife, which partially justified its inclusion of hold harmless provisions as an exercise of its public trust duties. Perhaps the public trust obligation really did inspire state officials to take on freshwater diversion at the expense of private oyster farmers. More likely, the justification was expressed belatedly in more recent leases as an anticipatory defense to legal challenge. Either way, the example shows how the public trust can be used early and proactively by the nonjudicial branches in order to improve natural infrastructure.

The second use of the public trust involved the judiciary. Here the court invoked the public trust to expand its view of state authority by characterizing the "redistribution of . . . productive oyster beds" as part of the state's obligation to restore the coast. This use is becoming more common. For instance, in *Parker v. New Hanover County* a North Carolina court upheld against constitutional challenge the state's power to level special assessments against private landowners in order to fund an inlet relocation project and restore damaged coastline.[142] Like the court in *Avenal*, the *Parker* court based its ruling on statute, constitutional obligations of stewardship, and the public trust doctrine.[143] Taken together, the court said, these laws urge "proactive steps to protect property from hurricanes and other storms," a lesson that was "brought home particularly keenly by recent hurricanes and their devastating impact along the Gulf Coast of the United States."[144] The public trust has also been invoked by courts to buttress government authority to improve lake quality in Louisiana, amend instream flow standards in Hawai'i, and ban personal watercraft in a Washington county.[145]

Third, at least one justice invoked the public trust as a defense against a takings claim by framing the state's diversion project as an exercise of police power, which, when combined with the public trust rationale, must be read to include the power to avert public "calamity." The idea is that because the state's duty to preserve natural infrastructure predates the property owner's title, that title must be limited by that obligation. States have long used the public trust duty as a shield against takings claims.[146] More than 30 years ago, in *Just v. Marinette*, Wisconsin's supreme court invoked its trust doctrine to help defend state regulation of private wetlands against a takings claim.[147] Embracing a "more sophisticated" understanding of ecology, the court found that the state's trust interest in traditional water resources gave it a trust interest in wetlands, too. "What makes this case different from most condemnation or police power zoning cases," the court wrote, "is the interrelationship of the wetlands, the swamps and the natural environment of shorelands to the purity of the water and to such natural resources as navigation, fishing, and scenic beauty."[148]

Cases like *Avenal* and *Marinette* show how the public trust doctrine can be used to protect public access to natural infrastructure. The strategy is not to expand the philosophical bounds of the doctrine, which continues to apply to traditional water resources, but to use modern science to expand the scope of what traditional water resources actually need. While laying the groundwork, these cases do not take the argument as far as it should eventually go. *Avenal* and *Marinette* were, after all, "friendly cases," as far as public officials were concerned, providing legal justification and protection for officials already disposed toward protecting infrastructure. The courts in these cases were not asked to *compel* public officials to act in environmentally responsible ways, nor were they asked to *rescind* anti-environmental decisions made previously. *Illinois Central* suggests that courts have these "unfriendly" powers, too. But I am not aware that any court has used the principles of ecological connection to *compel* or *rescind* management decisions affecting the public trust.

Doubters may argue that this absence is a good thing. Resource management affects private property interests, and it should be made as predictable and reliable as possible. When public officials make the decision to protect a wetland or divert river sediment, they necessarily engage in a public process that is different from a judicial hearing. Owners of private property should be shielded from the less predictable actions of "ecologist-judges." My response to this is that it is a mistake to see such disputes as affecting property interests on only one side. In disputes over natural storm barriers or fish nurseries, there are legitimate property interests on *both* sides: the private owner on whose property the infrastructure is located and the public user whose health, safety, or livelihood depends on the services naturally provided. As for the argument that traditional doctrines like the public trust or nuisance should always proceed in predictable ways, this view ignores the primary virtue of the common law system: its flexibility and ability to grow as common understandings change.

The point was emphasized in *Tenn v. 889 Associates, Ltd.*, the New Hampshire case that found a neighbor could use nuisance law to keep a neighbor from blocking his sunlight. Back in 1985, New Hampshire Supreme Court justice David Souter (later a U.S. Supreme Court justice) warned against a static reading of the common law: "If we were so to limit the ability of the common law to grow we would in effect be rejecting one of the wise assumptions underlying the traditional law of nuisance: that we cannot anticipate at any one time the variety of predicaments in which protection of property interests or redress for their violation will be justifiable."[149] Justice Souter was speaking about nuisance, but the point extends to other traditional property doctrines, including, I would urge, the public trust doctrine. To put it another way, property rights are more than prepackaged bundles of "powers and limitations," like the cartoonish characters in a deck of Pokémon cards. They are descriptions of *relationships* among private owners, neighbors, and the public. Like all relationships, the terms can evolve as knowledge and circumstances change.

To continue with a game metaphor, think about chess. Most people learning chess begin by studying the powers and limitations of each piece on the board. Based on the mobility of each piece, students are encouraged to assign "values" to their plastic soldiers in order to help them judge the efficacy of potential exchanges. A bishop or knight is worth "three," for instance, a rook is "five," and so on. While helpful in the beginning, expert chess players think little about this.[150] Instead, experts see the board in "chunks," that is, relationships among groups of pieces on nearby squares. In the expert view, a piece is not just a piece but an embodiment of its utility *as expanded or limited by the forces of all other pieces on the board.* The power of a free-range knight aided by an attentive pawn is worlds apart from a thumb-twirling bishop planted to protect the queen. What's more, these powers, like the relationships on which they are based, change all the time, so that the power or utility of a piece is best viewed in terms of its potential over a range of future scenarios rather than according to any static, bright-line rule. Property law will not and should not ever be *this* fluid. But the chess master's focus on relational utility can help us conceptualize a fairer and more productive way of resolving property disputes, particularly where public access to ecosystem services is at stake.

II

BE FAIR

5

BACKWATER BLUES

Whose tents are these?

—AL YOUNG

In the spring of 1927, in the state of Mississippi, people thought the rain would never stop. It resembled not the "the vivid brief downpours of April and May" but "the slow steady gray rain of November and December before a cold north wind." By this time the river had already overflowed its banks and submerged miles of farmland in several states. Water lay everywhere. Viewed from a bridge, the surface "was perfectly motionless, perfectly flat." But below that still layer "came a deep faint subaquean rumble which . . . sounded like a subway train passing far beneath the street, and which suggested a terrific secret speed."[1] Thus opens William Faulkner's "Old Man," the story of a convict sent to rescue a pregnant farm woman who is trapped in a tree and surrounded by water. The piece is justly celebrated for its structure and emotional restraint, balancing terror and vaudeville antics on the edge of a very sharp knife.[2] (Critic Malcolm Cowley famously described "Old Man" as the only other story about the Mississippi River "that can be set beside *Huckleberry Finn* without shriveling under the comparison.")[3]

For me, the genius behind this account of the 1927 flood lies in Faulkner's decision to see catastrophe through the eyes of the underclass. Chained prisoners, black sharecroppers, and stranded white farmers all navigate a world in which human beings are imprisoned not just by natural forces but by corrupt and narrow-hearted human institutions. Faulkner's perspective on the flood was not entirely new. The evacuation of thousands of homeless sharecroppers and shoeless children was already inspiring dozens of imagined and eyewitness musical accounts in the fast-developing American art form the Delta Blues, including Bessie Smith's "Back Water Blues," Charlie

Patton's "High Water Everywhere," and Alice Pearson's "Greenville Levee Blues."[4] What lasting power these artists tapped into! By the 1970s, thanks to a Led Zeppelin cover, even suburban white kids were crooning in thin falsetto to Memphis Minnie's 1929 classic "When the Levee Breaks."[5]

Faulkner and blues singers were not the only artists to court disaster. In the 1930s author and folklorist Zora Neale Hurston made Florida's Okechobee Hurricane the setting for the climax of her novel *Their Eyes Were Watching God*, which examines the hardships of Janie Crawford, a rural black woman blessed with determination but bridled by bigotry and community small-mindedness.[6] (Hurston herself had survived a violent hurricane in the Bahamas in 1929.) In 1939, John Steinbeck used the Dust Bowl storms of his generation to examine the broader themes of familial loss, economic oppression, and grassroots activism.[7] Steinbeck and Hurston present such hazards as naturally occurring, but as in "Old Man," the real tragedies are executed by humans against other humans—products of fear, bigotry, and malign neglect.[8] Disaster does not *create* such unfairness. Rather it heightens and accentuates bad habits already ingrained. That "subaquean rumble" you hear is not just the current of the river, it's the acceleration of social dysfunction.

As in music and literature, catastrophe in the real world follows a similar pattern. The heaviest burdens are borne by those with the least power—those who, for whatever social and economic reasons, are more exposed, more susceptible, and less resilient when disaster strikes. Groups on the higher rungs, squeezed by limited resources and fueled by fear, fall into old habits of distrust and control. Social structures designed to protect people from discrimination fracture under the mounting stress. Catastrophe is bad for everyone. But it is especially bad for the weak and the disenfranchised.

This view is not intuitive. Many people see events like earthquakes and floods as "social equalizers" whose wrath pays no attention to race, creed, or color. Some disasters are even portrayed as "community enhancers." In 1906 journalists used the term "earthquake love" to describe the sense of crosscultural unity believed to have developed after San Francisco's devastating earthquake and fire.[9] A modern version of the sentiment is captured on those television reports of seasonal floods in which all members of the community—men, women, blacks, and whites—are seen hoisting sandbags, distributing water bottles, and pulling wet dogs into rowboats. The phenomenon is real, as anyone who has lived through a major disaster can attest. But history shows that such effects are often short-lived.

This chapter illustrates the relationship between social unfairness and catastrophe, using a few well-known examples from America's past and two more current examples from Asia. Across time and across borders, naturally triggered disasters are nearly always accompanied by patterns of unfair social distribution. These patterns, which are rooted in bias, fear, and government neglect, unfairly subject some affected populations to more danger and hardship than others. In addition to the obvious injustice involved, these inequalities drive wedges between affected communities and inhibit recovery efforts.

Catastrophe and Unfairness in the United States

San Francisco's 1906 earthquake and fire marked one of the worst urban disasters in American history. The event decimated 90 percent of the city, killed an estimated 3,000–5,000 people, and plunged the region into near anarchy for weeks.[10] The transient and the working poor were particularly hurt. Many of them lived in hotels in the area south of Market Street, which had been built on the former Mission Bay Swamp. These buildings fell like dominoes when the quake hit.[11] Relief stations focused on areas of north of Market Street where the city's middle-class residents congregated. Official relief workers were scarcer in the Mission District and other poor areas. As historian Philip Fradkin notes, "poor white refugees who did not seek housing in the official camps received little help."[12]

Racism was acute, directed mainly toward residents of Chinese and Japanese descent. The vast majority of Chinese residents fled to Oakland and other nearby cities and were later forbidden to reenter San Francisco. The few hundred who stayed were held as "virtual prisoners" in relief camps on the Presidio.[13] During that time white residents looted Chinatown, sometimes under the supervision (and encouragement) of law enforcement officers. Many civic leaders resented Chinatown, which was wedged neatly between the city's financial center and an elite residential neighborhood. After the disaster they tried to move Chinatown south of San Francisco, but that effort was blocked by intense resistance from the local Chinese community. In Peking, the dowager empress herself voiced support for the San Francisco Chinese and threatened a trade war with her nation if Chinatown was not rebuilt in its previous location.[14] Those of Japanese descent were fewer. Some tried, unsuccessfully, to melt into the crowd; others left. Acts of violence against Japanese Americans, including stonings, were not uncommon. Such repressive actions, including school segregation, "at one point threatened to lead to a war between the United States and Japan."[15]

Women also suffered in particularized ways. Those without husbands or close relatives risked slipping into poverty in a postdisaster economy where traditionally female jobs had disappeared overnight. A report in the *Los Angeles Daily Times* described thousands of "forlorn and destitute women and girls . . . huddled in unhealthy camps."[16] As one female activist put it: "San Francisco is [now] a place for men, not for women. There is work of a manual character there in plenty, but the women are helpless. The means of earning a liv[e]lihood have been cut off completely. Girls who were employed in offices find themselves without even a prospect of work. Hundreds who had homes have lost their all, and many are without protectors."[17] Such conditions also had the effect of uniting women and catalyzing change. Historian Andrea Davies Henderson of Stanford University writes that disaster relief efforts placed more responsibility on the shoulders of middle-class female relief workers. Women's social clubs were transformed into charitable organizations. Evacuees in some camps became activists, demanding better living conditions for themselves and others. After she was denied a request to move her family from a leaking tent to a more suitable "relief cottage," activist Mary Kelly led a group of women to occupy a row of empty cottages, putting locks on the doors. Relief administrators had Kelly's house removed and destroyed—with her in it. After surviving this trial, Mary Kelly later fought for women's suffrage.[18]

In April 1927, the great Mississippi River overflowed its banks from Illinois to the Gulf of Mexico. For six weeks more than 20,000 square miles lay submerged beneath as much as 30 feet of water. An estimated 500,000 to 1,000,000 people lost their homes, and nearly 1,000 lost their lives. The human reaction to this horrific event, which played out in newspaper headlines across the country, abolished all pretense of southern gentility. In Mississippi, black men were hunted through floodwaters and conscripted into "levee gangs." In compensation, they received coarse food and mud-floored tents to sleep in. Faulkner's account matches the historical record when he describes scenes of "mudsplashed white men with the inevitable shot-guns" and "antlike lines of Negroes carrying sandbags, slipping and crawling up the steep face of the revetment to hurl their futile ammunition into the face of a flood and return for more."[19] Steamers sent to rescue white residents left black sharecroppers behind. (Planters feared that sharecroppers, once evacuated, might never return.)

Accounts of the evacuation of Greenville, Mississippi, are particularly vivid. There "[w]hite women and children massed around gangplanks waiting to board the steamboats" while barges carried "Negroes and terrified livestock" to concentration camps. "The National Guard patrolled the perimeter of the levee camp with rifles and fixed bayonets. To enter or leave, one needed a pass."[20] Daniel Farber notes that "[b]lacks and whites remaining in Greenville lived very different lives."[21] "About four thousand whites remained living on second floors, offices, or hotels. . . . In the meantime, about five thousand blacks were jammed into warehouses, oil mills, and stories; while over ten thousand more lived on top of the levees in tents with thousands of live stock."[22] Humiliated and betrayed by this experience, tens of thousands of blacks eventually left the Delta in search of better prospects in the drier, snow-swept North.[23]

Many poor whites were also victimized during the storm. White prisoners and homeless transients were also deployed in the notorious levee gangs. One particularly appalling event occurred as the flood rumbled toward New Orleans. Concerned that the city's levees would fail, New Orleans's banking elite persuaded the Army Corp of Engineers to relieve the river's pressure by blowing up a levee at Caernarvon, south of the city. The blast sent river water rushing at 250,000 cubic feet per second into the poorer rural parishes nearby, destroying farms and rendering thousands of people immediately homeless. City leaders had promised just compensation, but the process took years, and few victims ever received anything. New Orleans was spared, but some theorize that it would have survived untouched anyway.[24] To this day, the events at Caernarvon fuel speculation among some in New Orleans's black community that the levees were purposely dynamited during Katrina to flood the Ninth Ward and reduce harm to wealthier parts of the city.

A year after the 1927 flood, a lethal hurricane hit Florida, killing 2,500 people, "mostly poor blacks who drowned in the vegetable fields of the Everglades."[25] Perhaps no single day in history has ever seen the deaths of more African Americans.[26] The storm ruptured the dikes on Lake Okechobee and pushed a wall of water more than 10 feet high across miles of lowlands. As Hurston would later write, "[t]he monstropolous beast had left its bed" and was now "rolling the dikes, rolling the houses, rolling the people in the houses along with other timbers."[27] The victims Hurston describes were mostly black. As the storm approached the lake, some whites managed to evacuate; but according to journalist Michael Grunwald, "most blacks had to ride it out in their unprotected shanties in the low-lying fields."[28] In part, that was because the sole evacuation route was a winding two-lane road, and few black farmworkers had access to a car.[29]

On the subject of hurricanes, it was Hurston who wrote that "[c]ommon danger made common friends."[30] But the author was referring to the relationship between terror-stricken humans and terror-stricken wild animals, neither of whom appeared to notice the other while in flight. The relationship between terror-stricken humans and other terror-stricken humans was more complicated. Fearing the spread of disease, state officials quickly rounded up black male survivors and forced them at gunpoint to clear debris and bury the dead. Coffins were used only for whites (timber was scarce), while the corpses of nearly 700 African Americans "were stacked like cordwood on flatbed trucks and hauled to a mass grave in West Palm Beach."[31] This grave was eventually forgotten, covered by buildings and a street. Seventy years passed before the city officials finally purchased the land, put up a fence, and appropriately marked the site.[32]

As one last example of an American disaster, let's return to the San Francisco Bay Area and consider the more recent Loma Prieta earthquake of 1989 (magnitude 6.9). Though it was not nearly as strong as the one that brought San Francisco to its knees in 1906 (magnitude 7.8), damage was extensive in San Francisco, Oakland, and many other communities. The quake killed 67 people, injured 3,757, and left more than 8,000 homeless.[33] People of all stations in life were affected, but as in the past, the poor faced special challenges. As Ted Steinberg writes: "The media focused obsessively on [San Francisco's affluent Marina District], showing residents in Docksiders hauling their belongings about in plastic trash bags. But in truth, the worst hit area in the city was precisely the same one flattened in 1906: the south of Market area that was still home to skid row."[34] Indeed, an estimated 4,700 units of multifamily housing collapsed during the Loma Prieta earthquake, "precipitating a major housing crisis among the poor."[35] Ten years later, statistics showed that less than half of the affordable housing lost in the quake had been replaced.[36]

In Watsonville, a Mexican American community about 90 miles south of San Francisco, the damage was perhaps even worse. In the old sections of town, hundreds of wood-frame houses and apartments were reduced to rubble. Residents complained often that federal and state aid programs did not consider their needs. Most could not qualify for Small Business Association loans; FEMA's relief process was too complex. Some residents avoided government assistance out of fear that their immigration status, or that of family members, would be questioned. Thus Latinos in Watsonville remained homeless for longer periods than any other ethnic group affected by the earthquake. In 1993, it was estimated that "one-third of Watsonville Latinos lived in poverty exacerbated by the earthquake and recent cannery closures."[37] For the most part, residents of Watsonville remained invisible in the television news—a fact lamented by poet and essayist Al Young, who would later become California's official poet laureate: "[W]ho's going to cry or lose sleep," he wondered, "over a spaced out, tar papered, toppled / down town / by the sea, brown now with alien debris?"[38]

It is important to appreciate the tragic experience of the Chinese community in postquake San Francisco, of black farmworkers in the flooded South, and of Latinos in towns like Watsonville. But it is equally important to acknowledge that calamity can sometimes generate sparks of positive transition that when tended can erupt into noteworthy change. Such events never compensate for loss or injustice but serve to remind us that even in the darkest clouds can be found slivers of light. Thus the 1906 earthquake and fire forged alliances among female activists that later contributed to the suffrage movement and other progressive initiatives. The 1927 flood, which hastened the northward migration of black farmworkers, indirectly helped end the South's cruel sharecropping traditions. The political struggles that resulted from the Loma Prieta earthquake eventually led to a more equitable distribution of public resources for Latinos in Watsonville. In elections that followed that disaster, Latinos gained city council seats, and residents installed a Latino mayor.[39]

Catastrophe and Unfairness in Developing Countries

People in the developing world are least able to cope with disasters because of substandard infrastructure, poor emergency response, and insufficient medical services. Research suggests that the number of deaths from a natural disaster are generally linked to a nation's level of economic welfare and its degree of income inequality.[40] The world's 50 poorest countries, for instance, sustain just 11 percent of all natural hazards, but they suffer 53 percent of the world's hazard-related fatalities. In contrast, the world's richest countries experience 15 percent of all hazards but account for less than 2 percent of such deaths. The disparity remains even when the population density of rich and poor countries is taken into account. Because the economies of poor countries are generally less diverse and less resilient, long-term development can take a brutal hit after a catastrophic storm or quake. Oxfam International reports that "2 to 6 percent of South Asia's gross domestic product is lost to disaster each year."[41]

Inequality doesn't stop there. Within a country's borders, the relatively poor and the socially excluded endure special burdens. As in industrialized countries, poverty and bigotry push settlements into cheap and dangerous areas like floodplains or landfills. The grass huts and tin-roofed shanties that tens of millions of people call home provide little protection. Disaster aggravates engrained patterns of discriminatory treatment. In southern India, after the Asian Tsunami, human rights advocates reported widespread discrimination against lower-caste *dalits*, in the distribution of supplies, the removal of bodies, and the availability of shelters.[42]

Women and girls also face disproportionate harm. Poring over international surveys, one confronts the alarming fact that in the developing world, women and girls are much more likely to die during naturally triggered disasters. The Asian Tsunami claimed the lives of twice as many women than men in two hard-hit districts in the Indian state Tamil Nadu. In one district of Sri Lanka, women and girls accounted for 80 percent of the fatalities. In both the 2005 Kashmir Earthquake and India's 1993 Latur Earthquake, many more women died than men. The disparity is attributable to many factors. As primary caregivers, women are more likely to be home during a disaster, and their homes are often poorly constructed or in vulnerable areas. Women are usually the first ones to search for missing family members, exposing themselves to hammering rains, mud slides, and other perils. Sometimes women are constrained by social norms. During the Asian Tsunami many women drowned because they were "ashamed" to run to shore after waves ripped away parts of their clothing.[43] In some cultures, women and girls are not taught to swim.

On May 3, 2008, Cyclone Nargis tore through the Irawaddy delta in Myanmar, a low-lying coastal nation roughly the size of Texas. The delta is laced with hundreds of rivers and tributaries and rarely rises higher than 10 feet above sea level. Communities living on those muddy flats were pulverized. The United Nations estimates that 2.4 million people were severely affected by the cyclone and that 1.4 million people were left in immediate need of food, water, shelter, and medical care. Weeks after the storm, local officials estimated that 134,000 people had died or were missing, a figure that comports with early estimates by the Red Cross.[44] It is possible that the death toll was much higher, as Myanmar is run by a military dictatorship that is notorious for secrecy and a myriad of human rights abuses.

The events of the storm, which played out in news reports around the world, offer a shocking glimpse of yet another category of people especially vulnerable to natural hazards: citizens of oppressive, nondemocratic regimes that simply don't care. Myanmar's military government, for instance, had never bothered to draft a disaster relief plan for a major cyclone, although the threat was well known. And even after India's meteorological department notified the country's officials of the cyclone's approach two days in advance, the government did little to warn its citizens on the Irrawaddy.[45] In the aftermath, the government denied access to international aid workers and frustrated attempts to deliver supplies into the region, at one point leading the United States to accuse the government of "criminal neglect."[46]

There were also powerful aspects of ethnicity and class in this story that were nearly ignored in America's mainstream media. Myanmar has more than 100 different ethnic groups and subgroups within its borders, making it one of the most ethnically diverse populations in Southeast Asia.[47] The largest minority group, the Karens, live mainly on the Irrawaddy Delta where the cyclone hit. The Karens have suffered persecution from the governing junta since the country's early days of independence more than a half century ago.

After the storm, there were many reports of government officials hoarding supplies or selling them to victims at inflated prices. Some journalists described soldiers forcing survivors into hard labor, compelling them to "break up large boulders into pieces of rock for road construction" for less than a dollar a day.[48] Those in the area reported that Karens, in particular, were being discriminated against. Human rights workers cited unconfirmed reports that boatmen helping with the evacuation were taking ethnic Burmese from the delta but leaving Karens behind. Such reports are consistent with a historic pattern of discrimination against Karens by the Burmese government that includes violent attacks, property seizures, and forced labor. Some human rights observers believed such discrimination was "really the root of the [aid distribution] problem."[49] This situation can be seen as an extension of political and ethnic oppression that had been occurring for decades, just as violence toward African Americans during the Great Mississippi Flood of 1927 can be seen as an extension of generations of continued government-supported racial oppression.

Discriminatory abuse appears to have spilled over to other ethnic groups on the delta, including the Mon, Muslim Indians, aboriginal groups, and even the ethnic Burmese. A relatively wealthy class of Chinese landowners on the delta feared the government would use the evacuation as an opportunity to seize their land.[50] Hardship borne by many groups affects some more acutely because of past discrimination and other violations of human rights. For instance, millions of storm survivors were left homeless and denied interim shelter by even their own government. The homeless crisis in Myanmar thus equals the scale of the Asian Tsunami in 2004. But the tragedy is compounded by the fact that before the cyclone, an estimated 1 million people—most of them members of ethnic minorities—were *already* internally displaced, their homes having been seized or destroyed by soldiers in "counterinsurgency" efforts or to make way for dams, oil pipelines, and mining operations. In the half century before the storm, some delta communities had been relocated more than 100 times—often at the point of a gun.[51]

Children were particularly harmed. According to UNICEF's estimate, 40 percent of those who died in the cyclone and accompanying floods were children. Those children lucky enough to survive faced another set of struggles. Because of small body size, many were at special risk for disease and starvation. Children are also at much higher risk of protracted trauma and psychological problems, especially those who lose parents or other family members. Some independent relief centers, like those operated by UNICEF, provided "child-friendly" programs in which children played and made drawings as a means of coping with loss and trauma.[52] Hundreds, perhaps even thousands, of children were separated from their parents during the storm or, perhaps, orphaned. They now face the prospect of living in dour state-run orphanages or, worse, being swept up in the global slave trade. Myanmar has long been a "source country" for human trafficking, including domestic servitude, bonded labor, and forced prostitution. Less than two weeks after the cyclone, the United Nations had already reported an incident of two "brokers" entering a relief center in Yangon (also known as Rangoon) and trying to recruit orphaned children. The pair was arrested, but in the wake of the storm, the trend of child abductions and trafficking appeared likely to increase.[53]

The earthquake that shook China's Sichuan province in 2008 occurred in the same month as Cyclone Nargis. In contrast to that painful fiasco, Chinese leaders generally won praise for the efficiency and relative transparency of their relief efforts. Still, one can find evidence of distributional unfairness. As in Myanmar—or, for that matter, the United States—social problems that exist before a catastrophe are thrown into sharp relief when all hell breaks loose.

At a magnitude of 7.9, the Sichuan Earthquake was the strongest earthquake China had seen in 30 years. It killed more than 70,000 people and left almost 5 million homeless.[54] After the quake, tours of affected communities revealed that the poor had suffered most. In some poorer villages, where homes are made of cheap material and families live on $100 per month, nearly every building lay in ruins. As is common, most residents in these communities lacked insurance. Although China's central government set aside $772 million for relief aid, many villagers reasonably feared there would not be enough to go around. "The government has never helped us," said one villager to an American news service. "We don't trust that they will help us now."[55] Before the earthquake, government statistics showed that 2.1 million people in Sichuan province lived in poverty. Weeks afterward, the Chinese government reported that an additional 1.4 million farmers in 4,000 villages had fallen into poverty because of the destruction.[56]

In Beijing political circles, class is a sensitive topic. President Hu Jintao has based much of his public legitimacy on efforts to shrink China's widening economic gap, which is most noticeable between urban and rural populations. Days after the earthquake, an article in the *Wall Street Journal* contrasted two areas in the strike zone. In one, a rural village called Yinhua, "[b]oulders loosed by [the] quake, some as big as vans, littered the main road . . . along with the vehicles they [had] knocked over or crushed."[57] Nearly every home was destroyed. Meanwhile, roughly 55 miles away in Sichuan's capital city, Chengdu, conditions were very different. "The glitzy new office towers and hotels . . . were still standing and largely intact. The city suffered relatively little in [the] quake, despite its proximity to the epicenter."[58]

The collapse of nearly 7,000 school buildings in Sichuan brought more attention to China's economic inequality. At least 10,000 children perished in school buildings that many believe were poorly constructed and out of compliance with building codes. Thousands of enraged parents demanded government inquiries and protested what they called "tofu buildings." Many parents were "especially upset that some schools for poor students crumbled into rubble even though government offices and more elite schools not far away survived the May 12 quake largely intact."[59] Public outcry over such disparities appeared to knock the Chinese leadership on its heels. They were, after all, in the midst of preparing for Beijing's much-anticipated Olympic Games, which, would prove, with its "Water Cube" stadium and other stylish venues, to be one of the most expensive, lavish celebrations the world had ever seen. Weeks after the quake, Chinese officials were seen buying the silence of angry parents who had lost children in the calamity, offering them anywhere from $5,600 to $8,800 in cash in exchange for a signed promise to avoid protests and "maintain social order."[60]

It is perhaps inevitable that in human community there will always be social competition, a ladder of power that Hurston once likened to "the pecking-order in a chicken yard."[61] One of the jobs of government is to insist that within the chicken yard, all members have access to resources necessary for liberty under conditions that preserve a basic level of equality. These guarantees are hard to maintain in the best of times. When catastrophe hits, the importance of equal access to basic needs is magnified. But so are the threats against it. In the next chapter, I will show how racism and other forms of discrimination help fortify these threats, and I'll begin asking what can be done to address the problem.

Environmental Justice

Historians, relief workers, novelists, and blues singers have long understood the connection between catastrophe and social vulnerability. But legal scholars have rarely explored that territory. The study of law tends to reflect the organization of law itself, which in the American system resembles a vast field of supersilos. There is a silo crammed with rules about emergency response, federal aid, and other traditional aspects of federal disaster relief. Another silo holds several metric tons of environmental regulation and natural resource law. Another contains more than a century's worth of civil rights codes and initiatives for the poor. But the contents aren't mixed together as much as they should be. The same might be said for the discipline of law itself, which only recently has begun to mingle seriously with important disciplines like geography, behavioral psychology, and the earth sciences.

In Part I, I proposed that we look at natural resources in terms of their ability to reduce the risk of natural hazard. That examination required us to consider environmental and natural resources law in a larger context, so I sought to pull in the exciting field of ecological economics and a social theory of property to advocate for an environmental policy focused on natural infrastructure. That perspective, I believe, improves traditional environmental thinking by emphasizing holistic approaches to resource management (joining the needs of ecosystems and human community) and by strengthening the rights of the public to establish a reasonable interest over natural services provided on private property. In Part II, I want to stretch environmental and land-use policy in another direction, this time to include the needs of the socially vulnerable when faced with disaster. It may be true that few people, as Al Young suggests, lose sleep over a marginalized "spaced out, tar papered, toppled / down town."[62] Today the same is likely true of New Orleans, and of Galveston, too. But it should not be that way. Law should be reshaped to pay special attention to the poor and the marginalized during both disaster and recovery. To make my case, I must again draw from other fields of study, particularly geography and antidiscrimination theory. As it happens, these topics have already been put to good use in a legal and social movement known as environmental justice, which is concerned with the distribution of environmental harms and benefits on the basis of race, income, and other personal characteristics.

People may be created equal, but their patterns of pollution are not. Smog is rare in the country and common in the city. Dumpsites, often scarce in the suburbs, rise abundantly in the slums. The dirty river that loops safely around *your* neighborhood might strangle another. As for the "good" parts of the environment—the parks, the ponds, the tree-lined drives—these, too, emerge and recede in uneven patterns. The uneven distribution of environmental harms and benefits in the United States often means unequal environmental protection for whole classes of people living on the wrong side of the ecological tracks. A large body of evidence accumulated since the 1980s shows, for instance, that the poor and people of color are disproportionately subjected to hazardous waste facilities, air pollutants, contaminated fish, and pesticides. The concern for such inequalities, when they are based on income, race, age, sex, or other personal traits, is called "environmental justice."

Spawned by grassroots activists in the 1980s, the environmental justice movement first gained national attention as a protest against the placement of waste facilities and polluting industries in poor communities or communities of color. The agenda quickly expanded to address inequalities in other areas, including environmental health standards, worker safety, and government agency and political decision-making. When America's modern environmental movement began, most people did not know much about the distributional effects of environmental policy or—to be honest—care. Environmental protection was almost exclusively concerned with defining and controlling pollution levels *in the aggregate*, with little regard for distributional patterns.

Environmental justice issues span a broad spectrum, as do the activists involved. Environmental justice advocates come from the civil rights movement, the antitoxics movement, Native American organizations, the labor movement, and the traditional environmental movement. In the United States, the majority of leaders and activists in the movement are women.[63] Environmental justice efforts also exist in many other countries—both rich and poor—having evolved from myriad indigenous political movements. In 2002, when Wangari Maathai won the Nobel Peace Prize for her work with Kenya's Greenbelt Movement, some journalists had difficulty capturing the variety of her political goals, which emphasized government transparency, human rights, feminism, farm conservation, and tree-planting campaigns.[64] But her work could comfortably be contained in the term "environmental justice." Environmental justice advocates have also moved into international politics, most notably in the area of climate change, building a framework for what is called "climate justice."

In its American incarnation, the environmental justice framework embraces a concept of positive liberty and equality that is reminiscent of Joseph Singer's "social relations approach" to property law (discussed in Chapter 3). That is, every person should have equal access to some minimum level of resources so as to allow the pursuit of a safe, purposeful, and dignified existence. More specifically, the framework is generally rooted in a few widely shared principles. The first is that the "environment" should be interpreted broadly as including everything from wilderness to big-city streets to the factory floor. Environmental protection in this sense concerns any place where people live, learn, work, worship, or play.

A second principle is that all persons are entitled to protection from environmental harm. Advocates ground this right in the Civil Rights Act of 1964, the Fair Housing Act of 1968, the Voting Rights Act of 1965, "and even the 1948 United Nations Universal Declaration of Human Rights, which recognizes that people everywhere have intrinsic rights to life and health, and to a healthy environment."[65] This protection should favor prevention of harm over restoration or medical treatment after any harm and should place the burden on developers and polluters to prove the safety of what they propose. In addition, government agencies should give special attention to poor communities, communities of color, and other populations that bear disproportionate risks.[66] These views reflect environmentalism's traditional appeal to precaution, while emphasizing the needs of society's most vulnerable members—a traditional concern of civil rights and human rights advocates.

Finally, environmental justice advocates embrace the principle that communities speak for themselves and be heeded, insisting that those who are most affected by the pollution should have a central voice in the regulatory process" and encouraging community participation, neighborhood autonomy, and democratic decision-making.

While the diverse and overlapping nature of environmental inequalities makes categorization difficult, we might divide environmental justice problems into three kinds: inequalities associated with geography, inequalities associated with "one size fits all" health standards, and inequalities associated with legal and political processes. The most familiar of these is geographic location. Reports of polluted, inner-city "poverty pockets" or contaminated Indian reservations now regularly capture headlines in the local paper. But before that, it was Oprah Winfrey who, in the early 1990s, introduced America to "Cancer Alley," Louisiana, a string of small towns between Baton Rouge and New Orleans poisoned by hundreds of chemical plants and inhabited almost entirely by African Americans.

In addition to the anecdotes, a substantial body of evidence now documents the alarming extent to which polluting facilities are located in poor communities and communities of color.[67] One particularly influential report released by the Commission for Racial Justice of the United Church of Christ in 1987 found that communities with the most commercial hazardous waste facilities also had the highest population of racial and ethnic residents. While residents' socioeconomic status was an indicator for the presence of hazardous waste sites, the report concluded that race was the most significant indicator of all.[68] A follow-up study sponsored by the church ten years later revealed that despite more public awareness of the problem, not much has changed. "People of color and of low socioeconomic status are still disproportionately impacted and are particularly concentrated in neighborhoods . . . with the greatest number of facilities," the authors wrote, adding that "[r]ace continues to be an independent predictor of where hazardous wastes are located."[69] A study of air toxins in American cities conducted in 2006 found "a persistent relationship between increasing levels of racial-ethnic segregation and increased estimated cancer risk."[70] According to one account, over 57 percent of whites, 65 percent of African Americans, and 80 percent of Hispanics live in 437 counties with substandard air quality.[71]

There is no doubt that America's environmental laws have dramatically improved the quality of the nation's air, water, and wilderness, saving millions of lives along the way. But these efforts continue to produce radically unequal results on the basis of race, class, sex, and age. To give a few of many examples, African Americans have the highest asthma rates of any racial or ethnic group and are three times as likely as whites to be hospitalized for asthma treatment.[72] A similar pattern exists regarding lead poisoning. Poor children and children of color are eight times more likely to have elevated blood lead levels than other children.[73] In California, to use one example, 70 percent of children with lead poisoning are Latino, although Latinos make up only 45 percent of the young children in the state.[74]

The cleanup of contaminated properties raises other environmental justice issues. The divestment and blight that accompanies areas with more than their share of contaminated sites leaves poor people and people of color in a difficult situation. The first problem is that these areas have to compete with other contaminated sites for government cleanup resources. A 1992 *National Law Journal* report supports the claims of environmental justice advocates that sites in such communities are often neglected or receive less effective cleanups than sites in wealthier, predominantly white areas.[75] These differences in treatment and exposure are amplified by the lack of health services available to many vulnerable populations. Over 42.6 million Americans, including 10 million children, are without health insurance today. The uninsured rate for African Americans, Latino Americans, and Native Americans is more than one and a half times the rate for white Americans.[76]

On the opposite side of the geography problem, some neighborhoods and communities enjoy more environmental amenities than others, and the larger share of these lucky communities appear to be affluent and white. In Los Angeles there are 1.7 acres of parkland per 1,000 residents in areas that are disproportionately white and relatively wealthy, compared to 0.3 acres of parkland per 1,000 residents in the more racially diverse inner city.[77] In New York City, greenways, marinas, and swimmable water bodies are all more common in predominately white areas. In the city's five boroughs, not a single designated fish and wildlife habitat area is located outside a white community.[78] In addition, the government's emphasis on highway systems over public transportation disadvantages the poor and people of color, who are less likely to own cars and therefore more likely to rely on public transportation.[79] Access to usable transportation also favors affluent whites. According to a 2003 study cosponsored by the Harvard Civil Rights Project, only 7 percent of white households are without cars, whereas 24 percent of African American households, 17 percent of Latino households, and 13 percent of Asian households are. In urban areas, African Americans and Latinos make up 54 percent of public transportation users. As if to add insult to injury, the main public transportation routes in some metropolitan areas do not adequately serve their poorer areas.

Another important environmental justice issue involves "one size fits all" health standards, which sometimes fail to protect entire classes of people. The aforementioned case of asthma illustrates the need for strong air standards to protect children, who are at special risk for respiratory illness. (The elderly are too.) Women are more susceptible to PCBs, dioxins, and other dangerous chemicals that accumulate in fatty tissue. Pollution also threatens women's capacity to bear and nurse healthy children. In a few cases, susceptibility to pollution varies physiologically by race. In addition, minority groups and the poor—who suffer disproportionately from anemia, heart disease, and low weight births—are more vulnerable to the cumulative effects of toxic exposure. Despite these important differences, many environmental health standards continue to be based on the vulnerabilities of a healthy white man between the ages of 25 and 35.[80]

Unfair health standards follow from inaccurate assumptions not only about who we are (all white, all male, all adult) but about how we live. Consider fishing. Federal water quality standards aspire to keep most water bodies safe for fishing and swimming. In setting acceptable levels for waterborne toxins, the EPA once assumed that the average person ate no more than one fish meal per month. But this average concealed dramatic variations. In the South, the rural poor rely heavily on subsistence fishing, as do Native Americans in the Northwest and some people of South Asian descent. Some tribal populations in the Puget Sound consume at least 10 times that "average" amount of fish. Today, thanks in part to the efforts of environmental justice advocates, the EPA now uses a separate consumption standard for subsistence fishers that is much higher than the norm. Still, more work remains to be done. The newer standards can only take into account consumption patterns of subgroups whose habits are known and documented. But most tribal and racial subgroups still have not been studied. Without more information about subsistence fishing habits throughout the country, environmental protection for these groups is incomplete.[81]

In developing standards to protect farmworkers from pesticides, agencies estimate exposure without recognizing that children, even infants, often accompany their parents in the fields, and that these young people are more vulnerable to pesticide toxicity. Because farmworkers in the United States are mostly Latino and almost always poor, these regulatory oversights result in further race-based and class-based inequalities.

From a broader perspective, the current chemical-specific approach to assessing environmental health risks fails to account for the reality that people of color and the poor are often exposed to several different pollutants at the same time. The cumulative and synergistic effects of multiple exposures can be dramatic. An effort to regulate toxic mixes began in the Clinton administration but has not moved forward.

In addition to physical differences in risk, people's *perceptions* of risk vary according to race and sex, suggesting that political decisions about acceptable health risks should incorporate the views of a wider range of the population. A national survey conducted by James Flynn, Paul Slovic, and C. K. Mertz found, for instance, that, even when correcting for differences in education and income, "white males tended to differ from everyone else in their attitudes and perceptions—on average, they perceived [environmental and nonenvironmental] risks as much smaller and much more acceptable than did other people."[82]

Finally, environmental justice advocates consistently raise issues about environmental decision-making and public participation. Environmental decision-makers traditionally have heard the views of industry giants, conventional environmental organizations, state and local governments, and federal land managers, but not the people who actually live in the areas most affected. Influence in environmental decision-making is a main priority of environmental justice organizations. As Deeohn Ferris, former executive director of the Washington Office on Environmental Justice, likes to say, "We want more than a place at the table, we want part of the *meal*." Yet even where access is allowed, meaningful participation can be difficult. Conventional stakeholders have significantly more time, money, and other resources to bring to these processes and influence agency policy and implementation. Environmental justice advocates often lack the resources to participate as effectively in such a highly technical arena, and this fundamentally tilted playing field significantly compounds the problems just explained, producing bad decisions that harm public health.

Because local environmental justice organizations operate on such slender budgets, they must rely heavily on publicly available information. They are thus unequally burdened when government restricts public information, as the Bush administration did after 9/11. Government claims for exemption under the Freedom of Information Act are practically routine in some agencies. The Homeland Security Act and the Information Quality Act (which allows individuals to challenge the release of government reports) also significantly limit many types of formerly public information relevant to environmental protection. Together these initiatives could allow firms to withhold information about hazardous emissions, accidents, and other risks posed by power plants, nuclear facilities, refineries, chemical plants, and other large facilities. In a democracy, information is power. This move toward greater secrecy can only further disempower our most environmentally vulnerable communities.

The distribution of environmental harms and benefits throughout society does not occur randomly. Many environmental justice advocates argue that race, sex, and other immutable characteristics affect one's likelihood of exposure and susceptibility to environmental threats. Some cite even intentional bias among corporate or governmental decision-makers. Defenders of the system often respond that distributional patterns result from market forces and consumer choices. The landfill, the smelter, the chemical plant do not come to poor or minority neighborhoods in order to harm residents, they come to take advantage of affordable land and friendly land-use regulations. In other cases, market dynamics cause poor or minority populations to congregate around existing facilities.

Whatever the cause, it is hardly fair for minorities and the poor to suffer such environmental harm when so many others do not. The unfairness mounts when one considers that it is the middle and upper classes whose consumption is most responsible for the country's extraordinary levels of pollution and waste. Aspiring to distributional equality would actually lead to better and more efficient environmentalism by internalizing the costs of consumption. If consumers of wasteful products shared more equally in bearing the environmental costs of production and disposal, they might encourage more ecofriendly products, more pollution prevention, and fewer and smaller landfills. Environmental justice activists, therefore, think of themselves not as NIMBYs ("Not in My Backyard") but as NOPEs

("Not on Planet Earth"). Finally, a greater emphasis on environmental costs would awaken privileged groups to environmental vulnerabilities that they *share* with the less privileged but are unaware of as yet. Such awareness, in turn, can lead to more forceful and unified campaigns toward environmental protection.

In promoting environmental justice, activists emphasize solutions that give local communities leadership in decision-making. Environmental justice struggles thus celebrate coalitions that take charge: a group of organizations in Emelle, Alabama (some black, some white) that fight for higher standards for the town's gigantic hazardous waste dump; the multiracial organization in Oakland that educates residents about lead poisoning; the Mothers of East Los Angeles who successfully marched against a proposed incinerator in their already blighted community. And at least some parts of the country (New York City and Los Angeles are prominent examples) appear to be in the midst of an "urban parks movement," where diverse coalitions of city dwellers are joining to encourage and protect parks and other environmental amenities. Environmental justice activists warn that grassroots activism cannot succeed by itself. In their view, government must play the leading role. This is true not only in a practical sense but also in a moral one; for the mission of government is to protect the welfare of citizens, *regardless* of income, race, or other characteristics.

The crosscutting nature of environmental justice problems suggests the need for a broad array of strategies in many fields. Most obviously, officials should avoid the circumstances that allow the many ills described here to take root. Local governments (and federal officials who sometimes oversee them) should not permit polluting factories or waste dumps in areas already plagued by such facilities. State and federal enforcers should investigate violations in the inner city just as aggressively as in the suburbs. Standards for toxic cleanup and brownfield development should remain consistent across lines of race and income.

California provides one example of such efforts. Its new city planning statute requires long-range development plans that equitably distribute public services and avoid "overconcentrations" of industrial facilities near schools and homes. According to the state's draft guidelines, "[o]verconcentration occurs when industrial facilities or users do not individually exceed acceptable regulatory standards for public health and safety, but when considered cumulatively with other industrial facilities and uses, pose a significant health and safety hazard to adjacent residential and school uses."[83]

There is work for the federal government, too. Advocates argue that the federal government should revise national transportation policy to mandate cleaner vehicles, reduce urban air pollution, and emphasize public transportation in minority and low-income communities. In addition, agencies responsible for setting health and safety standards should revise their methods of risk assessment to take into account differences among various subgroups and differences caused by cumulative and synergistic effects.

The Clinton-era Executive Order 12898 on Environmental Justice could provide a catalyst for all of these things. This 1994 Order, which remains in effect today, requires that all federal agencies "to the greatest extent practicable" pursue environmental justice goals.[84] Specifically, the Order directs agencies to do two things. First, they must identify and address disproportionately high and adverse health or environmental effects on minorities and low-income populations in their programs and activities. Second, they must ensure that future programs and activities that substantially affect the environment do not create unequal burdens on the basis of race, color, or national origin. The Order provides that each federal agency "shall make achieving environmental justice part of its mission by identifying and addressing, as appropriate, disproportionately high and adverse human health or environmental effects of its programs, policies, and activities on minority populations and low-income populations in the United States and its territories and possessions."[85] After years of neglect and poor focus, though, the promise of regulatory fairness under the Order has fizzled. In 2005, the Government Accountability Office found that the EPA had "generally devoted little attention to environmental mental justice" when drafting three significant clean air rules between 2000 and 2004.[86] In 2006 a report by the EPA's inspector general found that 60 percent of EPA's program and regional office directors had not performed environmental justice reviews as required in the Order.[87] Moreover, courts have held that the Order creates no right of judicial review.[88] Having grown frustrated with the Order's vague requirements and troublesome definitions, some environmental justice advocates are now arguing for an amended version.[89]

Because environmental justice so directly involves racial discrimination, one might look to federal civil rights laws for relief. But such laws, while important, have serious limitations—in part because they do not address the special circumstance of environmental issues and in part because a more conservative judiciary has, over the years, made such laws less robust. Consider the equal protection clause of the Fourteenth Amendment to the

Constitution. The clause forbids the government to, among other things, discriminate on the basis of race. Courts read this provision to bar only government conduct that *both* produces a discriminatory outcome *and* reflects a clear racist intent.[90] In practice, this almost always means that government laws or conduct must explicitly mention race to establish the intent requirement. But government conduct that abets environmental injustice—poor zoning or siting decisions, inadequate permit requirements—almost never meets this test. Thus, it is possible for a county to site nearly all or most of its landfills in minority communities without tripping the equal protection wire.

In contrast to the equal protection clause, some civil rights statutes allow plaintiffs to prove discrimination on the basis only of discriminatory outcome. But many of these statutes, which forbid discrimination in education, employment, or housing, are poor fits for most environmental claims. One statute at first presented a glimmer of hope. Title VI, part of a broad civil rights act passed in 1964, holds in section 601 that "[n]o person in the United States shall, on the ground of race, color, or national origin, be excluded from participation in, be denied the benefits of, or be subjected to discrimination under any program or activity receiving Federal financial assistance."[91] Section 602 authorizes federal agencies to pursue the statute's goals by issuing regulations that bar activities that have disparate impacts on protected groups.[92] But federal agencies and the courts have limited the full potential of Title VI.

On the regulatory side, the EPA has issued regulations under section 602 that forbid government-supported activities that produce disparate impacts. But the EPA's assessment of citizen complaints is far from adequate. Investigations by the EPA, which receives nearly all the Title VI complaints related to environmental justice, have crawled at a snail's pace in reviewing community allegations. The agency has also repeatedly failed to define key concepts in its regulations that are essential to proper review. From 1993 to 2003 (the last date for which data have been compiled), the agency received 136 Title VI complaints, yet only 16 have been accepted for investigation.[93] None to my knowledge has led to a formal finding of discrimination.

Given this backlog, the need for private rights of enforcement in the courts has increased. Unfortunately, the Supreme Court has severely limited the use of private litigation as a means of enforcing disparate-impact regulations issued under Title VI. In a recent decision involving a nonenvironmental, private lawsuit, the Court interpreted the act to provide a private remedy only for claims of *intentional* discrimination brought directly under section 601.[94] Read this way, Title VI is no more helpful to private litigants than the equal protection clause. Some lawyers are exploring a legal theory by which complainants could avoid the "intent" requirement by suing the government under the Civil Rights Act of 1871 for essentially the same claim. This theory—which involves technical readings of both statutes—was recently rejected by the federal Court of Appeals for the Third Circuit, although it might still succeed in other jurisdictions.[95] What seems clear is that if Title VI is to provide any protection from environmental discrimination in the courts, the "intent" requirement will probably have to be eliminated—by either a more sensible judicial interpretation or by an act of Congress.

The short history of the environmental justice movement offers insight for those interested in improving disaster policy. One insight helps better define the problem as a combination of external threats and personal or community vulnerability. A second insight suggests that distributional fairness is important for reasons of both fairness and political stability. A third insight suggests ways regulation or litigation might be used to protect the welfare of the least advantaged. We'll look at these ideas in Chapters 6 and 7. If American environmentalism was nurtured by the belief that, in Barry Commoner's words, "everything is connected to everything else," it is now time to admit another truth—that every*one* is connected to every*one* else, too.[96]

6

DISASTER JUSTICE

The social dynamics that result in unfair distributions of pollution and toxins play similar roles in the area of disaster management. Environmental justice as understood up to this point has involved what we might call chronic, slow-motion disasters in which damage can take years or decades to emerge. By contrast, disasters like storms and floods reveal what some researchers have called "environmental injustice in the fast-forward mode."[1] Such inequalities have become the concern of a growing number of experts and activists who might be described as advocates for "disaster justice."

Like environmental justice, disaster justice is about the failure of law—specifically, the failure of law to provide vulnerable people with the protections and benefits they need to lead safe and productive lives. To describe and address the legal failure, environmental justice advocates offer several insights that will help us in our examination of disaster policy. First, the environmental justice movement interprets the environment expansively to include forests as well as inner cities, bubbling streams as well as asbestos-lined pipes, the campground as well as one's home or workplace. Applying this idea to disasters means looking beyond the buckled fault line or the blazing tall grass to see within our definition of environment the crumbling apartment buildings and the soot-filled air, the crowded relief centers and the broken water and sewage lines.

A second insight from environmental justice can be expressed by an equation: risk = exposure × vulnerability. Environmental justice advocates know that risk cannot be separated from the victim's individualized exposure and vulnerability. Further, they know that both variables can be exaggerated by the forces of poverty and bias. Through civil rights laws, environmental laws and regulations, and private tort actions, environmental justice advocates attempt to use the legal structure to impose accountability on bad actors. When advocates submit evidence suggesting that African Americans are more likely to live near a hazardous waste landfill or that certain Indian tribes consume disproportionately more mercury-laden fish, they are illustrating a link between race and increased *exposure* to an environmental hazard.[?] Not surprisingly, the definition of exposure has become one of the most controversial issues concerning the EPA's disparate-impact regulations issued under Title VI.[3] Environmental justice advocates also pay close attention to the physical and social vulnerabilities of those who are exposed. People likely to have other toxins in their bodies through multiple exposures, for instance, are likely to be more vulnerable. The same is true for people who lack access to good preventative health care or for children and the elderly, whose respiratory or immune systems may not be as robust.

Third, environmental justice advocates see the protection of public health, safety, and the environment as being "public goods," that is, benefits to which everyone to some degree is entitled and that cannot always be justly or efficiently allocated through the private market.

Finally, the environmental justice movement encourages us to look for patterns in the social problems we study. Establishing patterns is, in fact, an important strategy in many social movements, including the civil rights struggle and feminism. For instance, one barrier women faced in characterizing sexual harassment or domestic violence as "institutional" issues was that they lacked information about the prevalence and similarity of such occurrences. It was not until the "consciousness-raising" era of the 1970s that researchers began gathering significant volumes of information about sexual harassment, domestic abuse, divorce settlements, child custody disputes, and the like. Until then, it was possible for critics to dismiss calls for protection as based on unique or unusual cases that did not represent society as a whole. Connecting the dots was important. Environmental justice advocates experienced a similar awakening in the 1990s, when for the first time, national and regional studies from many sources began confirming a pattern of disproportionate environmental harm based on race and class. Such studies eventually led to many state laws and a federal executive order recognizing the institutional nature of environmental injustice.[4]

This insight is important in a book that uses Hurricane Katrina as a source of broader lessons. There is a temptation to dismiss New Orleans as an anomaly, a circumstance that, no matter how tragic, has little to do with the rest of the country or the world. The view is understandable. New Orleanians have spent almost 300 years trying to convince outsiders that they live in one of the most unusual and elaborately nonconforming places on earth. There are unique problems in New Orleans. and we do have our own particular dysfunctions. But many of the dysfunctions I will discuss exist in many American cities. It is vital to see this connection, because the misfortunes of New Orleans are part of a national, institutional pattern involving race, gender, and environmental protection. If we dismiss these warning signs as exotic novelties, we will have wasted a chance to improve disaster policy in the United States. Indeed, the lesson of Chapter 5 is that for people living on the margins, disasters are more similar than they are different. These events, across many generations and time zones, all share a pattern. The pattern has to do with social and economic power. But the pattern can be shaped through law. And it is the health of the law as it exists in peaceful times that gives an indication of the degree of justice one can expect in times of disaster.

Developing a theory of disaster justice will require us to examine some of the social, economic, and political issues that are familiar to environmental justice advocates. I will, for instance, unpack the concept of vulnerability to understand the origins of some of the special risks some people face in times of catastrophe. The law's failure to protect vulnerable groups in such times represents, I will argue, a fraying of the "social contract," a set of government responsibilities intuitively understood by most Americans. This failure follows in part from the free-market economic model that has hollowed out government services and encouraged a "you're on your own" attitude. This failure is also perpetuated by intentional and unintentional biases (racism, sexism, and the like) that keep vulnerable groups from building resilience into their communities. Having rendered this diagnosis, in Chapter 7 I will offer some tentative legislative and regulatory prescriptions.

Exposure and Vulnerability

In New Orleans, many of Katrina's effects—the demolished homes, shattered communities, and lost jobs—were borne disproportionately by people of color, the poor, and women. (Reported deaths were nearly proportional to the city's racial demographics but were distinguished by age: more than 60 percent of those who died were elderly.) In contrast to the conventional view of disasters as "social equalizers" or "community enhancers" (discussed in Chapter 5), Katrina showed once more that naturally triggered disasters follow both demographic and geographic patterns. The reason has to do with what disaster researches refer to as "exposure" and "vulnerability."

EXPOSURE

Exposure refers to the mostly physical aspects of a disaster that put people in harm's way. New Orleans, a subtropical city at the mouth of a deltaic river, is hurricane bait, as are Houston, Miami, and even Washington, D.C. The course of the Atlantic hurricane belt, although driven mainly by nonhuman forces, nonetheless has a racial character: for historical and geographic reasons that involve the antebellum slave economy and racist policies thereafter, a large majority of American counties or parishes that are disproportionately black lie near the Gulf or the south Atlantic seaboard—within striking distance of an angry hurricane.[5] In the three Gulf states hit hardest by Katrina—Louisiana, Mississippi, and Alabama—African Americans represented more than a quarter of the population.[6] This exposure means that national hurricane policy is not only a Gulf Coast issue or an Atlantic Coast issue but a specifically African-American issue—that is, an issue in which African Americans should be particularly interested and involved.

Within the city of New Orleans, geography also plays a role in disparate exposure. Although New Orleans was (and is) more racially and economically integrated than many American cities, African Americans (regardless of class) and the poor are more likely to live in parts of the city that are more prone to flooding, because of elevation, levee configuration, or pumping networks. Damaged areas in the metropolitan area were 45.8 percent African-American, while undamaged areas were only 26.4 percent African-American.[7] In New Orleans proper, the damaged areas were 75 percent African-American, while undamaged areas were 46.2 percent African-American.[8] Thus, two of the most devastated areas—New Orleans East and the Lower Ninth Ward—were almost all of color and were notoriously prone to floods and storms. Such housing patterns, of course, did not occur by chance but followed formal and informal segregation efforts, as well as traditional market forces. As any native Orleanian will tell you, "Water flows away from money." Even so, thousands of families living in mostly affluent or mostly white communities also suffered extreme damage, including residents of the more upscale suburb of Lakeview and the nearly all-white St. Bernard Parish, whose structures were obliterated by storm waters barreling through the Mississippi River Gulf Outlet (MR-GO).[9]

Race continues to be a potent variable in explaining the spatial layout of urban areas, including housing patterns, street and highway configurations, commercial development, and the siting of industrial facilities.[10] The differential residential amenities and land uses assigned to black and white residential areas cannot be explained by class alone. For example, poor whites and poor blacks do not have the same opportunities to vote with their feet. Racial barriers to education, employment, and housing reduce mobility options available to the black underclass and the black middle class. On the Gulf Coast, race-based housing patterns developed after the Civil War when whites chose areas for black settlement that were burdened by flooding, unhealthy air, noise, and poor infrastructure. In New Orleans, black neighborhoods were typically shunted toward "low-value, flood-prone swamps at the end of the city."[11] The pattern continued through the 1980s, when many middle-class and lower-middle-class blacks moved to the vulnerable eastern parts of the city, attracted by inexpensive apartment complexes along an elevated portion of Interstate 10.

When minority communities were not moving toward flood hazards, the hazards were moving toward them. In the early 1900s, construction of New Orleans's Industrial Canal cut through the Ninth Ward, isolating the mostly "lower" section (which was predominantly black) and exposing it to the danger of levee breaks.[12] In 1958, the Army Corps of Engineers located MR-GO, sometimes referred to as the city's "Hurricane Highway," St. Bernard Parish and the Lower Ninth's northern border. That project significantly degraded the area's natural protective infrastructure. For instance, MR-GO is estimated to have caused the loss of 27,000 acres of wetland storm buffers due to erosion.[13] Levees along that waterway and Lake Pontchartrain have profoundly weakened soils throughout the city; and portions of the Upper and Lower Ninth Ward have subsided up to 10 inches in the last 40 years.[14] In 2009 a groundbreaking decision from a federal court in Louisiana found the Army Corps liable for its negligent maintenance of MR-GO. That failure, the court found, was directly responsible for flood damage in St. Bernard Parish and the Lower Ninth Ward. The ruling could pave the way for future lawsuits resulting in tens of billions of dollars in damages, or, perhaps, a government compensation package established by Congress. The case, *In re Canal Breaches Consolidated Litigation*, is almost certainly headed for appeal and should be watched closely.[15]

Many disasters indirectly expose residents to dangerous chemicals, gases, or toxins. In New Orleans, Katrina's floodwaters inundated at least three Superfund sites (Bayou Bonfouca, Madisonville Creosote Works, and Agriculture Street Landfill), each located near poor African-American communities in New Orleans. The extent to which the flooding exposed residents to increased levels of toxins is still being debated.[16] The flood also covered other sites in some predominantly white areas, including the working-class community of Chalmette. There, the flood surge slammed into an oil refinery and ruptured a storage tank, releasing more than a million gallons of mixed crude oil into residential neighborhoods. A class-action lawsuit brought by 6,000 homeowners against the refinery owner, Murphy Oil, was recently settled for $330 million.[17] Events like these raise serious concerns about the geographic distribution of dangerous structures like landfills and refineries, as well as concerns about how they are secured and maintained. Here environmental justice concerns and disaster justice concerns begin to overlap. Minority communities in New Orleans have fought battles for years over the location and maintenance of heavily polluting facilities. The primary motivation has been to minimize the chronic, slow-motion threats like increased birth defects and long-term illness. But Katrina showed that the same lax regulation that increases the risk of a slow leak can also set the stage for sudden collapse.

Once contaminants have been released into the urban environment, it should be the responsibility of industry and the government to remediate the damage and stop the exposure. But after Katrina, even that mission proved controversial, I later found out. Katrina left a range of environmental problems in her wake, including flooded sewage plants and oil and chemical spills. But the mud had barely dried on the streets when the city's mayor, Ray Nagin, reopened the city for business and rehabilitation. "I know New Orleanians," he chimed, "and once the beignets are in the oven, once the gumbo is in the pot . . . they'll come back."[18] According to early environmental reports, it was true that the large municipal area, as a whole, had been spared the "toxic stew" scenario in which the city proper became a cauldron of industrial poisons, raw sewage, and venomous snakes. Certain communities were clearly at risk, particularly those affected by oil spills or sludge from the canal bottoms. Officials with the EPA and the Centers for Disease Control voiced concern about whether the city's immediate reoccupation plan was wise. But no federal agency ever directly challenged the city's judgment. When it came to public health, the "beignet and gumbo" standard was all we had.

Around this time, toxicologist Wilma Subra had begun taking test samples of the soil and water in various parts of the municipal area. A generation ago, Subra's efforts to protect residents living in Cancer Alley had earned her a MacArthur Foundation "genius" grant. Studies of Subra's soil samples taken just after the storm documented three neighborhoods in New Orleans where levels of arsenic, benzo(a)pyrene, and petroleum hydrocarbons exceeded state and federal residential standards.[19]

Shortly after the storm, I visited one of these neighborhoods in the central part of the city near the Industrial Canal. The area, a low-rise apartment complex, had been swamped with canal waters when the levees burst. Everything there was in disarray. Trees were shattered and street signs torn down. Abandoned cars lay beached on the sidewalk like dying whales. The streets and sidewalks were filled with broken furniture, molding rugs, and soaked appliances. Outside the door of one unit, a wheelchair leaned gently toward the sun, planted in a foot of mud. I met two women, a mother and daughter, picking their way toward the entrance of one of the buildings. The daughter carried a large flathead shovel in one hand and a plastic sponge mop in the other.

We began a conversation. This was their first trip back into the city. It was my second, I said. Friends had told them there was little left of the mother's ground-floor apartment, but they were hoping to salvage a few pieces of china. They seemed remarkably cheerful for the task at hand, if a little unprepared. Experts were encouraging residents to wear special protective suits and N-95 respirators, but few people I knew could find such things at the time. In contrast, the daughter was dressed in faded jeans and a windbreaker, her brown face circumscribed by a tightly cinched, bright yellow hood. The mother wore a bandana around her neck, a purple sweat suit, and a pair of white shrimper's boots. Having mentioned I was an environmental law professor, I asked them if they knew that the soil and mud around the mother's home was contaminated with bacteria, mold, and arsenic—the last in amounts up to 10 times the acceptable levels. They seemed concerned, but laughed it off. "If that don't get you, something else will," said the mother.

Another disaster-related environmental justice issue involves debris removal. Virtually any environmental calamity will require a speedy and large-scale cleanup. After Katrina, it is estimated that 22 million tons (or 55 million cubic yards) of debris was removed and disposed of in Louisiana alone.[20] In the 1980s, some unlined landfills in New Orleans had been closed when contaminants were found leaching into the groundwater. But after Katrina, many of those landfills were reopened to dispose of disaster debris. One study estimated that the wood debris might contain as much as 1,740 metric tons of arsenic, which if disposed in an unlined facility could threaten area groundwater.[21] Many of these landfills, of course, are located near poor or minority communities. For instance, a Vietnamese American community in East New Orleans remains involved in a dispute to protect their neighborhood and a nearby wildlife refuge from potential hazards posed by a local landfill, which was allowed to accept potentially toxic storm debris without first installing a leachate collection liner. As one health policy institute's 2008 report puts it, "catastrophic damage inevitably leads to dramatic increases in demand for solid waste disposal, and chaotic conditions frequently limit opportunities to effectively sort hazardous from non-hazardous debris. Under these conditions, the likelihood remains high that minority and low-income neighborhoods will be burdened disproportionately with water and air pollution from debris removal and burning, given the historic pattern of siting landfills in those areas."[22] Solid waste disposal, long a centerpiece of many early environmental justice struggles, is also important for advocates of disaster justice.

There is also sometimes a gendered dimension to exposure in times of disaster. Before Katrina, New Orleans had (and continues to have) a very high proportion of single mothers, many living in poverty. Their lack of resources and dependence on a network of family and neighbors for childcare conspired to keep them in neighborhoods more prone to flooding in the years before the storm. More immediately, the same factors made them less mobile as Hurricane Katrina approached and evacuation warnings went out. The large majority of people left stranded in the Superdome and New Orleans Convention Center—with insufficient food, water, and medical attention—were African-American women and children, many of whom had no means to flee the city and no other place to go. This point brings us to the topic of vulnerability.

VULNERABILITY

Exposure cannot completely explain the storm's uneven effects. Recall that after the levees broke, a full 80 percent of the city lay underwater, including some prominent, mostly white neighborhoods. The storm and the levee failures lumped greater harms on the poor, people of color, and many women, because as a group they were more vulnerable. Experts who study disaster know the importance of social vulnerability, which they define as "the characteristics of a person or group in terms of their capacity to anticipate, cope with, resist, and recover from the impact of a natural hazard."[23] Geographer Susan Cutter notes that "[s]ocial vulnerability is partially a product of social inequalities — those social factors and forces that create the susceptibility of various groups to harm, and in turn affect their ability to respond, and bounce back (resilience) after the disaster."[24]

AFRICAN AMERICANS AND THE POOR

For generations, New Orleans has been the poster child for social vulnerability. Before the 2005 flood, 28 percent of people in New Orleans lived in poverty.[25] Of those, 84 percent were African American.[26] Of people five years and older living in New Orleans, 23 percent were disabled.[27] An estimated 15,000 to 17,000 area residents were already homeless.[28] These vulnerabilities play out in a number of ways, affecting everything from evacuations to long-term rebuilding efforts.

The dysfunctional evacuation during Katrina is now legendary. Although 2 million coastal residents managed to flee, thousands of people stayed behind — some by choice, others for lack of reasonable options. Of the 28 percent of New Orleans households who lived in poverty, many had no access to a car: 21,787 of these households without a car were black; 2,606 were white.[29] This lack of access, of course, became crucial, given an evacuation plan premised on the ability of people to get in their cars and drive out of New Orleans. Days later, these stranded residents were eventually delivered to — or as BBC News described it, "unceremoniously dumped" in — cities throughout the nation; and they were not provided the means to later return home.[30] Hundreds of thousands of people, many of them poor and African American, are believed to have been "permanently displaced," making the event America's largest involuntary migration since the Dust Bowl, perhaps larger.

Poor people and people of color also tend to suffer more psychological effects from disaster than victims who are wealthier or white.[31] These vulnerable groups are also less likely to have access to mental health services. Higher rates of stress and psychological trauma are rooted, not surprisingly, in a more general lack of financial and political control of one's circumstances. Elderly blacks, in particular, have been found to recover more slowly from "psychosocial" trauma than whites, an effect partially attributable to financial security.[32] Fortunately, poor people and people of color are able to compensate somewhat by relying on the strength of what sociologists call "social capital," the support derived from networks of extended family and friends. Thus in one study of a tornado in rural Mississippi, the black children appeared to rebound more quickly than the white children.

The former profited not only from a closer family relationship but also from a structured schedule planned around household responsibilities and farm chores.[33] Part of the tragedy with Katrina was that the flood destroyed so much, sweeping away whole networks of families, churches, and friends. As for structure, the storm left thousands without homes, schools, jobs, or plans of any kind. Much has been written about the violence and chaos that fell on New Orleans in the days after the flood. But perhaps the bleaker story is found in the weeks and months of purposeless frustration and sheer boredom that so many endured as they *waited*—in shelters, food lines, medical clinics, bank lobbies, legal aid bureaus, and job pickup sites—just *waited* for some sign, however trivial, that their luck would someday change.

In addition to the limitations of poverty, disability, immobility, stress, and trauma, African Americans found their options constrained by the old straitjacket of racism. Sensationalized rumors of black rioting and looting, according to many anecdotal accounts, made local white residents and some enforcement officers less trustful of African Americans calling for help. The epitome of such behavior is captured in the dramatic standoff on the "Bridge to Gretna."[34] That bridge, also called the Crescent City Connection, is a segment of highway connecting New Orleans to the city of Gretna, a predominately white, blue-collar suburb of about 17,000 people. When New Orleans filled with water, that bridge span was one of the very few ways out of the city. Three days after Katrina struck, about 200 evacuees left the wretched conditions of the Superdome and Convention Center and set out toward the Crescent City Connection in hopes of reaching food, water, and shelter. When this group tried to cross the bridge, they were met by a line of armed Gretna police officers. The officers fired their shotguns into the air and told the party to turn back because the city had been "closed." Nearly all of the evacuees on the bridge were African American; the police officers were all white.

We may never know what thoughts were going through the minds of the Gretna police, but eyewitnesses (most of whom were evacuees) said the police were motivated by race-based fears. "I believe it was racism," said one evacuee to a British journalist. "It was callousness, it was cruelty."[35] Gretna officials denied that charge. They said the bridge was closed only when lawlessness began erupting in nearby communities. "If we had opened the bridge," Police Chief Arthur Lawson told a reporter, "our city would have looked like New Orleans does now, looted, burned, and pillaged."[36] The city of Gretna remains in a tangle of litigation over the events of that day;

and the incident continues to generate outrage among social justice advocates in the United States and abroad. In fact, however, similar events have often played out in disaster narratives. The officers firing warning shots over the Gretna bridge are not unlike the steamboat captains who refused to take any black passengers during the 1927 Mississippi River Flood or the boatmen in Myanmar who bravely rescued their Burmese kin but reportedly ignored the pleas of ethnic Karens (Chapter 5). Perhaps what is surprising about Gretna is that such callousness could take place in twenty-first-century America. But if disasters teach us anything, it is that a people should never consider itself immune.

Today, years after Katrina, redevelopment and city planning have also been affected by race and class issues. Louisiana used millions of dollars in special federal Community Development Block Grants and tax credits to develop programs to rebuild the state's housing stock. These included two programs to promote the repairing of large and small rental properties and one program, the "Road Home" program, to compensate homeowners for the costs of rebuilding. While these programs greatly accelerated the rebuilding process, they have been criticized for not doing enough for the city's vulnerable populations. Because of inadequate funding, the rental repair programs have only supported the repair of fewer than one-third of the 82,000 rental units lost to Hurricanes Katrina and Rita.[37] As for homeowners, nearly three-quarters of Road Home applicants had significant gaps between the received rebuilding resources and the actual costs of repair. The average shortfall for African Americans ($39,082) was roughly $8,000 more than it was for whites ($30,863).[38] In the Lower Ninth Ward, which was almost entirely African American, the gap between a household's average damage estimate and the average government grant exceeded $75,000.[39] The inequality resulted from the basing of the formula for grant awards on a home's prestorm value rather than the actual cost of repair. Because homes in African-American neighborhoods had lower prestorm values (on average), their owners received less, even when their repair costs were similar to those in other neighborhoods.

Rebuilding after a major disaster also requires revised city planning and large-scale public development, but such projects raise understandable skepticism among communities of color. (Recall the proposals to "redevelop" San Francisco's Chinatown a century ago.) In 2005, for example, the nonprofit Urban Land Institute unveiled a master plan for New Orleans that would have encouraged denser development on natural ridges and reserved some lower areas for wetlands and flood control. But the proposal, which failed to involve local communities and which provided no plans for housing those in reclaimed areas, was soon dropped. For now, the only firm plans involve 17 projects, from schools to community centers, that the city's redevelopment agency hopes will encourage further development. As of this writing, a master plan is being developed for the city by a Boston architecture and planning firm that will have the force of law when finalized. This plan is a subject of much debate in minority communities, as some fear it will seek to turn "black space" into "green space," without adequate minority involvement and without meaningful plans to relocate neighborhoods located in vulnerable areas.

WOMEN AND CHILDREN

From San Francisco's 1906 earthquake to the 2004 Asian Tsunami, we have seen how women bear special burdens in times of catastrophe (Chapter 5). Katrina was no exception. As is true in most American cities, women have made up the majority of the working poor in New Orleans.[40] Still, New Orleans has had an especially high index of gender-based economic segregation. In New Orleans, lower-income women care for children, bathe the hospital patients, and clean the hotel rooms; lower-income men drive the trucks, patch the roofs, and fill the potholes. Single mothers are also much more likely than single fathers to be the primary custodians of their children and to live in public housing. Indeed, 88 percent of all households in the city's public housing units were headed by single mothers.[41]

Given these factors, women were hit particularly hard by Katrina. Of the 180,000 Louisianans who lost their jobs after the storm, 103,000, or 57 percent, were women. Of the thousands of households that lost public housing when New Orleans' housing authority closed four projects after the storm, 88 percent were headed by women. Men's median annual income rose after the storm, in part due to the rise in heavy-labor jobs like demolition and construction. Women, who were more likely to work in the health-care, education, and hospitality sectors, saw their median income decline. Such widespread destruction, of course, dramatically increased stress within families, predictably leading to soaring reports of domestic violence. In the weeks after the storm, reports of rape and other crimes against women also increased. Indeed, research shows that evacuations and disasters are often accompanied by increases in violence against women and girls.[42]

Women bear more subtle burdens, too. More than a year after Katrina, tens of thousands of families in Orleans Parish were still picking up the pieces. They were navigating insurance and public grant programs, supervising their contractors, ordering new appliances, organizing community recovery efforts, enrolling their children in new schools, finding new doctors, caring for traumatized grandparents, and negotiating living space with an assortment of well-meaning but still homeless relatives. The family members most involved in these duties tended to be women. Many women quit paid jobs or reduced their work hours to take on these frustrating and time-consuming responsibilities.

The uneven burdens experienced by African Americans and women contributed to a shift that occurred in the city's demographics following the flood. Soon after the flood, Nagin famously called for a return to New Orleans as a "chocolate city," by which he meant a city that was predominantly black.[43] Population estimates more than a year after the flood show that New Orleans morphed temporarily into a "mocha city," with a black plurality (about 46 percent) but no racial majority. By 2008 the city's African-American population had rebounded to nearly pre-Katrina levels (about 61 percent).[44] The city has also changed in terms of gender. Before Katrina, New Orleans had a narrow majority of women—it was a "Venus city," if you will.[45] It is now a city of "Mars," with men making up slightly more than half of the population.[46] In the first months after the flood, the city looked like one of those mining towns you see in Westerns, void of families and stocked with a menagerie of whiskered, unbathed men. Only weeks after the storm, I was naively surprised one Thursday night to find the French Quarter alive and hopping. By seven thirty, the bars on Bourbon Street were already packed, while a line of laborers and off-duty FEMA agents slouched toward a popular strip club. The demographics are more balanced now (and the lines on Bourbon Street, to my eye, are shorter); but as of 2006, two-thirds of the single mothers who left New Orleans near the time of the hurricane had not returned.[47] And it is unlikely that they ever will.

In addition, the lack of child care, pediatric services, and neighborhood schools—along with a short-term spike in mold-based allergens—made the city less hospitable to children. A study sponsored by Columbia University's Mailman School of Public Health in 2008 found that Hurricane Katrina's youngest survivors are now the sickest children in the nation.[48] As one illustration, the authors' review of medical records of children living in a federally funded trailer park showed that 41 percent of children younger than age four had iron deficiency anemia; 24 percent of all children there had respiratory, allergic, and skin ailments; and 55 percent of elementary-school-aged children had behavioral or learning problems. Such disorders are thought to be related to the environmental conditions in and around the FEMA trailers people are still living in, as well as the disruption of family patterns (including diet) and the stress of dislocation.

OTHER RACIAL OR ETHNIC GROUPS

The flood displaced nearly 40,000 Mexican citizens living in the New Orleans area.[49] Many Latinos in the New Orleans area could not speak English, and there were many complaints about "inadequate language capacities to deal with affected Latino residents" in the Gulf region.[50] Similar language problems arose during the Loma Prieta quake.[51] Many Latinos in Louisiana were undocumented workers, laboring on farms or in the "hospitality industry" (restaurants and hotels). As in the aftermath of the Loma Prieta earthquake, many Latinos avoided seeking disaster relief after Katrina out of fear of being deported, despite assurances to the contrary from U.S. immigration authorities and President Vincente Fox of Mexico.

At the same time, thousands of Latinos moved to the Gulf Coast states, where they diligently gutted houses, reroofed buildings, and poured many thousands of new concrete foundations. In New Orleans—a historic port city founded on multiculturalism—the new Latino presence blossomed everywhere. Radio stations like Radio Tropicale Caliente beefed up their offerings and expanded their audience. In supermarkets, the variety of tortillas, peppers, and salsas expanded overnight. Spanish-language churches, for example Metairie's El Buen Pastor Baptist Church, were soon busting at the seams. The parking lots at Lowe's and Home Depot became makeshift public squares, hosting dozens of Latinos who congregated to find construction work, exchange housing information, and snack on the fiery tamales dished out at mobile food stands.[52] Statewide, the growth of Hispanic populations far outpaced Louisiana's population growth as a whole.[53]

For the most part, residents have welcomed these new visitors who, after all, were the mainspring of the city's recovery. But stories of anti-Latin bigotry and actions perpetrated by whites or other racial groups in expression of such bigotry are not uncommon. The labor market has proved especially oppressive. Two independent studies have reported that laborers in New Orleans often report poor health and safety conditions at work, substandard housing, and getting stiffed on wages promised by employers.[54]

Thousands of Native Americans were also blasted by the storm. A representative of the National Congress of American Indians estimated that "there [were] several thousand Native Americans living in the hurricane's path" in Louisiana, Mississippi, and Alabama.[55] The bayous southwest of New Orleans are home to roughly 20,000 Native Americans, most of whom belong to tribes that are recognized by the state of Louisiana but not by the federal government. As one example, the Biloxi-Chitimacha tribes in southern Louisiana live near eroded and vulnerable marshland in subsistence-based tribal communities that were literally washed away by Hurricanes Katrina and Rita (the latter causing more damage in this area). Roughly 80 percent of the homes in the area lay submerged in several feet of water and mud. Federal aid was slow to reach them, in part because their lack of federal tribal recognition kept them off the political radar screen and (literally) off of government maps; FEMA officials later admitted that they did not even know that Indians lived in the lower bayous.[56]

In addition to illustrating vulnerability, this story also says something about exposure. The reason these tribes live so close to the water is that since the 1940s, oil and gas interests have been gradually edging them south (through suspect sale and lease agreements) to make way for drilling operations. And the reason the marshes are now so eroded and vulnerable is that these operations have sliced the lowlands into ragged strips of grass. Describing the situation in a radio interview, Marlene Forêt, who leads the Grand Caillou/Dulac Band of the Biloxi-Chitimacha, echoed the same theme of natural infrastructure we saw in Part I. The tribal community, she explains, "is more vulnerable now than it was way back. Oil companies came in and drilled for gas, oil, and the canals, in many of them—and the water kept coming in and saltwater, and it just killed the vegetation there, and then as the water was coming in, then the land could not replenish itself."[57]

The storm also affected many people of Asian descent. The bustling Vietnamese American community in eastern New Orleans was all but obliterated when a segment of levees collapsed to the south and water rushed in. Reflecting the strength of their community and religious networks, this neighborhood was one of the first to rebuild after the storm. In addition, nearly 50,000 people of Vietnamese descent labored on the Louisiana coast, most in the fishing and shrimping industries. Many of their homes and livelihoods were permanently damaged. The same fate was visited on hundreds of Filipino-American shrimpers, whose Filipino community in Louisiana is one of the oldest in North America.[58]

BEYOND KATRINA

The disaster injustice resulting from Katrina was extreme for modern times but not unique. Consider Galveston, Texas, which was devastated by Hurricane Ike in 2008. Before Ike, Galveston had about the same poverty rate as New Orleans (22–23 percent). Galveston's poor population, as in other parts of Texas, was disproportionately African American and Latino. One-third of the adult African-American population and slightly less than one-half of the Latino population did not have a high school diploma or the equivalent. Hurricane Ike (Category 2) ripped through about three-quarters of Galveston's housing stock. Especially affected where middle-income and low-income apartment buildings. Most of those apartments, according to a report in the *Houston Chronicle*, "were located among the city's most vulnerable neighborhoods, where blight already had driven many buildings into disrepair even before the storm."[59] Residents of Galveston's storm-battered public housing projects were ordered to leave the unsafe structures; no alternate arrangements provided for suitable housing.[60]

As with Katrina, Hurricane Ike also compromised facilities storing chemical toxins. Preliminary reports from the National Oceanic and Atmospheric Administration suggested that there were at least 40, and as many as 100, releases of hazardous materials in areas hit by the storm.[61] One photo released by this agency showed oil streaming from pipes into a flooded wetland near the city.

The prestorm evacuation efficiently cleared more than 2 million people from the coastal area, but uncertainties about immigration enforcement policies, as after Katrina, discouraged many such persons and relatives of such persons from taking advantage of transportation services. After the storm, when government officials were answering calls to repair water lines and help tarp roofs, many immigrants already living on the margins were afraid to request these services, out of fear of being arrested or harassed.[62]

Disaster justice is not only of concern to the southern United States, with its tropical storms and history of slavery. Think of New York City, the region ranked second in the world in terms of economic assets, threatened by coastal flooding and sea-level rise. Scientists believe that a Category 3 storm like the one that raked Long Island in 1938 could send a storm surge as high as 25 feet into some parts of New York City. City officials estimate that the homes of as many as 600,000 people could be flooded and 3 million people forced to evacuate, with economic losses topping $100 billion.[63]

Many of the effects of environmental and disaster injustice result from what social scientists call racial isolation and poverty concentration. New Orleans had a high degree of racial isolation before Katrina and still does. The level of black-white segregation in pre-Katrina New Orleans, for instance, was 69 (the national average is 60), meaning that about 69 percent of blacks would have had to move in order to achieve the same geographic representation as whites. Black poverty concentration describes the percentage of black poor people living in neighborhoods where 40 percent of the residents are impoverished. Using that measure, New Orleans scored a 31.2, much worse than the national average of 19. But New York City also has high numbers. Its rating for black-white segregation is 81.8; its rate of black concentration of poverty is 32.5.[64] Should a storm surge flood parts of the city, it is not hard to imagine the race-based and class-based inequalities that would result. Aging buildings in poor neighborhoods would collapse or catch fire. Industrial facilities located in minority communities might release harmful fuels or other toxins. Looting would occur, and loose bands of vigilantes would form to protect their neighborhoods, some bearing handguns, shotguns, even assault rifles, as in New Orleans.[65] An evacuation would be especially challenging, particularly if the subway lines flooded, which they are expected to do; 40 percent of New Yorkers do not own a car.[66]

Imagine a devastating earthquake in Los Angeles, where racial segregation is just as high as in pre-Katrina New Orleans (although the concentration of black poverty is not), where dozens of languages are spoken, and where the fears of immigration enforcement would drive thousands of victims away from the very government services their lives would depend on. Even in Los Angeles, 11 percent of residents do not own a car (just below 14 percent in New Orleans), making it difficult for them to leave dangerous areas or seek emergency care. One can play this grim scenario many times over—a flood in St. Louis, an earthquake in Oakland, an earthquake *and* flood in Sacramento—in cities across America where social vulnerabilities in the form of poverty, segregation, and disability would send the country's most needy residents into a spiral of continued want and despair.

A Broken Contract

More than two centuries ago, feminist Mary Wollstonecraft stormed, "[i]t is justice, not charity, that is wanting in the world!"[67] In times of disaster, charity in the United States abounds. Americans across the country organize food drives, donate blood, drive hundreds or even thousands of miles to contribute their skills and labor. But when we speak of the efforts of government—the deployment of rescue teams, the disposal of hazardous debris, the design of long-term investment and recovery programs—we are no longer talking about the kindness of strangers. We are talking about justice.

Social critic and Georgetown University professor Michael Eric Dyson suggests that the difference between charity and justice has something to do with reliability. "You know you get tired of giving," he said in an interview shortly after Katrina. "You get tired of generosity. You get fatigued of compassion, but structures of justice are more permanent. Justice continues the impulse that charity begins."[68]

Philosophers have often described justice in terms of a "contract" between a government and its citizens.[69] Journalist Michael Ignatieff used the justice-as-contract metaphor to explain the Katrina debacle. He argues that government's multisided failure—and here he includes, among other examples, the shoddy levees, the bungled emergency response, the awful conditions at the Superdome and the Convention Center—was more than bad luck or incompetence. It was a breach of the "contract of citizenship [which] defines the duties of care that public officials owe to the people of a democratic society." While admitting that the provisions of this contract are contested, he argues that at a minimum, government owes its people reasonable "protection," that is, a promise that government will "help[] citizens to protect their families and possessions from forces beyond their control." Ignatieff's analysis is worth quoting at length:

> When the levees broke, the contract of American citizenship failed. . . . The most striking feature of the catastrophe is not that the contract didn't hold. That is now too obvious to argue about. . . . What has not been noticed is that the people with the most articulate understanding of what the contract of American citizenship entails were the poor, abandoned, hungry people huddled in the stinking darkness of the New Orleans convention center. "We are American," a woman at the convention center proclaimed on television. . . . [T]hat single sentence was a lesson in political obligation. Black or white, rich or poor, Americans are not supposed to be strangers to one another. . . . Citizenship ties are not humanitarian, abstract or discretionary. . . . So it is not—as some commentators claimed—that the catastrophe laid

bare the deep inequalities of American society. These inequalities may have been news to some, but they were not news to the displaced people in the convention center and elsewhere. What was bitter news to them was that their claims of citizenship mattered so little to the institutions charged with their protection.[70]

Ignatieff concludes by predicting that "[m]illions of acts of common decency and bureaucratic courage will be necessary before all Americans feel that they live, once again, in a political community and not in a savage and lawless swamp."[71]

Ignatieff's essay highlights an important distinction between chronic social justice issues and the concerns of disaster justice. Good people can disagree about what role government should play in eradicating poverty or providing health care. We can even quarrel over the placement of a refinery or landfill in a poor neighborhood. But, really, there should be no disagreement about protecting the disadvantaged from cyclical storms or tinpot levees. For nearly 100 years, that has been as basic a duty as protecting our shores from military invasion.[72] That our leaders stripped such basic security from us—without any notice or even debate—is cause for alarm, and should be a source of deep shame.

Restoring the promise of personal security and shared citizenship should be what disaster justice is all about. America must extract itself from the "savage and lawless swamp" it has slipped into and make good on the promise of "political community." But first we must examine the slip. What circumstances gave rise to such quick abandonment of Louisiana's homeland security? What does that abandonment say about the state of government in America today and the prospects for improving it?

As in Part I, a little property theory helps frame the discussion. In my examination of natural infrastructure, I suggested that we look at disaster buffers like coastal wetlands and healthy forests as public goods. By that I meant that such features provided advantages to so wide a range of users that it was difficult to exclude anyone from the benefit without closing the benefit to all. That condition is called "nonexclusivity." In addition, public goods are so plentiful that the use by one individual does not usually detract from a similar use by another individual, a condition called "nonrivalry." Because of their nonexclusive and nonrival character, public goods are hard to privatize; therefore, I argued, their care is often best left to government. When public goods are successfully protected and maintained for all, they provide two desirable outcomes. First, they offer *on an equal basis* a set of fundamental goods and services (in this case security of persons and property) that promotes the *exercise of liberty and human flourishing* among users. That is the ethical, or political, payoff. Second, broad and equal access to the services of natural infrastructure creates synergistic spillover benefits that accrue to the public at large, even after individual users have had their share. That's the economic payoff.

Although they rarely talk about public goods, advocates for social justice in the United States have often framed their arguments in a way that evokes this strand of property theory. A century ago, tragedies like the 1911 Triangle Shirtwaist Factory fire (which killed 146 immigrant female workers locked in a burning building) and the Great Mississippi Flood of 1927 gave power to the progressive movement, which argued for a more active role for government in fulfilling citizens' basic needs.[73] The central argument was that basic safety (whether occupational or environmental) was a public service that all workers or residents were equally entitled to as a matter of common decency, or as Wollstonecraft might say, *justice*. In addition, progressives argued that government's provision of minimum safety would accrue benefits for the public at large by maintaining a secure and healthy workforce and (in the case of flood protection on the Mississippi) a reliable stream of commerce.

The feminist movement of the 1970s also put forward this idea. Kicking through the door separating "private" needs and "public" entitlements, activists suggested a new role for government in protecting the social and economically vulnerable. Many relationships that in more traditional times were seen as private transactions came to be seen, with the help of feminism, as more public. Thus, sexual harassment in the workplace is no longer a problem of "personal relations" to be negotiated individually by the parties but is seen today as an impermissible workplace condition. The same goes for domestic violence, once also seen as a private matter, not a social problem or a crime. A workplace free of harassment or a household free of violence is not something people have to bargain for anymore: they are *public entitlements*, guaranteed to all Americans and protected by the state. And their availability, as we are correctly told by advocates, makes *everyone* better off by helping to provide a more productive workforce and a more stable family structure.

Environmental justice advocates are perhaps explicit in drawing the connection between justice and public goods. And in doing so, they have helped move the ball forward. In a report published by the Sage Foundation after Hurricane Katrina, a group of influential environmental justice scholars, including Manuel Pastor, Robert Bullard, and Beverly Wright, encouraged their readers to think of environmental safety and risk reduction as "'impure' public goods."[74] They are helpfully viewed as public goods because, like workplace safety and a violence-free home life, they help individuals flourish while creating synergistic spillovers for all. They are "impure" because, as these authors point out, environmental safety and risk reduction are not entirely nonexclusive. That is, it is possible to protect *most* people from leaking landfills or surging waters while not entirely protecting *everyone*. "For example," they write, "flood-control projects provide location-specific benefits, restricted to those people who live or own assets in the protected area. By virtue of where they live, work, or own property, some members of society reap the benefits of such collective investments, and others do not."[75]

"Impure" public goods raise questions about allocation. Storm-buffering cypress swamps can be maintained in ways that protect everyone in range, including marginalized communities like those of the Biloxi-Chitimacha tribes. Or parts can be leased away to private concerns for use as canals or garden mulch. How should we decide how such benefits are distributed? We can imagine two options (with room for compromise in between): first, the *justice* approach, which underlies the environmental justice movement, and second, the *market* approach.

Based on my examination of environmental justice in Chapter 5, we can now contemplate how the justice approach would distribute the benefits of environmental security. When a minority is excluded from security that is available to the majority, there is both a moral and an economic shortcoming. The moral, civil shortcoming is the deprivation of positive liberty and equality. The economic shortcoming is the loss of public spillover effects. Environmental justice advocates have been especially effective in showing how expanding environmental protection to excluded minorities can translate into benefits for all. Controlling pesticide use in strawberry fields protects not only migrant workers but also upscale consumers. Restrictions on emissions from oil refineries not only protect minority fence-line communities but also reduce greenhouse gas emissions. Had government officials been better able to protect vulnerable people during the New Orleans flood, the city would have benefited in multiple ways, preserving for public benefit a more stable workforce, safer streets, and a more focused and organized army of community volunteers.

Drawing from the environmental justice movement, the new advocates of disaster justice argue that the allocation of safety and risk reduction should be governed by a process sensitive to the values of liberty, equality, and broad social benefit. That stands in contrast to the system we have now, which more closely resembles a market system in which consumer demand determines distributions of security and in which racial and economic disparities are more easily tolerated.

The Market-Based Approach

According to the market-based approach, hazard protection should be allocated the same way other goods and services are allocated in the marketplace. Protection should follow explicit or implicit market signals. The current resurgence in America of the market-based approach might accurately be said to have begun when President Ronald Reagan proclaimed in his first inaugural address: "Government is not the solution to our problem. Government is the problem."[76]

Reagan's approach to market solutions is grounded in an intellectual movement called neoliberalism, a revived form of traditional liberalism that champions free markets and individual liberty in an economy gone global.[77] As geographer David Harvey puts it, "[n]eoliberalism is in the first instance a theory of political economic proposes that human well-being can best be advanced by liberating entrepreneurial freedoms and skills within an institutional framework of strong private property rights, free markets and free trade."[78] Some may believe that neoliberalism as a guiding principle is waning in the Obama administration. This is not entirely so. While it is true that President Obama and, perhaps, the public have embraced a more optimistic view of government and its role in American life, current economic forces will ensure that American law and international law continue to follow a market approach to solving big problems.[79] Thus the goal, from a disaster justice perspective, is not to reverse the market approach (as it can't be done) but to make space within the neoliberal framework for a vocabulary of justice. In this area, the Obama administration may prove a helpful ally.

From the neoliberal model, three relevant corollaries follow. (We'll see more corollaries in Part III.) First, neoliberal policy seeks the efficient allocation of resources. "Efficient," here means optimizing aggregate social welfare in a context of limited political and material resources. "Efficient" does not always mean "fair," and for this reason free-market ideology is sometimes described as "amoral." Second, free markets are much better at allocating resources than are governments or other organized institutions. This is what Reagan meant by "Government is the problem." Third, neo-liberalism promotes an ethic—some would say "virtue"—of self-sufficiency and the stoic acceptance of unfortunate consequences. Individuals are expected to assume the risk of participating in the market, and to adapt quickly to changing landscapes. "Instances of inequality and glaring social injustice," in this view, "are morally acceptable, at least to the degree in which they could be seen as the result of freely made decisions."[80] Indeed, as neo-liberal philosopher Robert Nozick has argued, efforts to redistribute wealth in order to rehabilitate economic losers creates its own injustice by treating affluent individuals as a "means" to enhance the "ends" of those who are less affluent.[81]

The market approach poses a problem for advocates of disaster justice for a couple of reasons. First, as we saw in Part I, protection from disaster requires infrastructure, and infrastructure, because its benefits are shared, is hard to fund through private means. Second, justice (in the sense I use the term) requires attention to distributional outcome, and that value is not recognized in a purely market approach. The market approach has been especially hard on urban areas, where reliance on infrastructure is high and diversity is great. Geographer Stephen Graham notes that cities rely on "vast complexes of infrastructure, public works, and hazard mitigation systems," which we normally take for granted as part of the "public realm."[82] The neoliberal attack on "big government" resulted in a draining of large-scale public works and social safety net programs on which urban populations rely. Indeed, the most prominent urban policies of neoliberals at the federal level emphasize voluntary services provided by churches and other non-profits to help the needy. The result of these antiurban policies and fiscal shifts, according to Graham, "has been to make [metropolitan areas] intensely vulnerable to social and natural catastrophe, and to raise levels of violence and disorder.[83]

That statement pretty much explains the scene in New Orleans after the federal levees broke. Weeks later, President Bush was promising "bold action" to subdue the entrenched poverty of the Gulf states.[84] But his resolve did not last beyond the next news cycle. The problem was that bold action against poverty requires something far beyond what the small government approach can muster. Indeed, neoliberalism appears to have resulted in increased poverty and a much wider wealth gap. The same week that Katrina struck in 2005, a Census Bureau report was released showing that poverty had increased in America for the fourth straight year. According to the report, 5.4 million Americans had fallen below the poverty line, increasing the country's poor population from 11.3 to 12.7 percent of the whole.[85]

It sometimes seemed that conservatives were trying to lower any expectation that the government was capable of doing anything. John Tierney, then a regular columnist for the *New York Times*, argued for completely erasing the federal role in disaster response. "New Orleans and other coastal cities will never be safe if they go on relying on Washington for protection," he wrote.[86] Conceding the federal government's duty to replace New Orleans's flawed levee system, Tierney suggested that local residents should then rely only on private insurance and local response teams like fire departments. "Here's the bargain I'd offer New Orleans," he chirped; "the feds will spend the billions for your new levees, but then you're on your own."[87]

Disaster victims since Katrina appear to be reluctantly taking this message to heart. After the 2008 floods and levee breaks in Iowa that swamped nearly 2,000 homes in Rapid City, flood victims soon complained about conflicting information they were receiving from federal agencies on everything from new building standards to government buyouts. Financial uncertainty was paralyzing the city's rebuilding effort. Then New Orleans residents arrived and set up workshops explaining how to organize at the grassroots level and how to proceed without waiting for government aid or involvement. "When the New Orleans organizers showed up in town, we thought, 'These are our allies,'" said one resident. "The bottom-up strength that New Orleans exhibited has to be our model." The main piece of advice, described in a newspaper headline for this story: "Do it yourself."[88]

So yes, Americans are a practical, market-driven, self-reliant people. Nonetheless, Americans see a positive role for government, too. President Obama picked up on this sentiment and rode that wave to the White House. It was no coincidence that in the summer before his election, one survey showed that eight in ten Americans agreed with the following statement: "The social contract of the twentieth century, an agreement between the government, employers and society that affords Americans with basic necessities of the American Dream, appears to be unraveling."[89]

The most charitable critique of the market approach is that its followers are naive and misguided. Inspired by the myth of a self-correcting economy, market fundamentalists merrily send their fellow Americans into the woods to be devoured by wolves. More thoughtful market advocates will concede the danger of an unfettered market but insist that government intervention would make things even worse for everyone, including the poor. A less charitable—but still moderate—critique holds that market advocates know very well the wolves are out there, and probably think government intervention can successfully keep them at bay; but these advocates are unwilling, for selfish reasons, to adopt government protections because such protections would have the result of converting some private resources (enjoyed by the upper class) to public resources (enjoyed by everyone, including the lower class).

Canadian journalist Naomi Klein takes this critique one step further. In her book *The Shock Doctrine: The Rise of Disaster Capitalism*, Klein depicts "free-market" democracy as an intentionally exploitative and destructive force. In her view, the United States and other powerful nations have taken advantage of high-profile "shocks"—terror attacks, economic meltdowns, and natural disasters—to impose free-market solutions that end up raiding public resources, cutting wages, and silencing citizens. She calls this cutthroat strategy "disaster capitalism." The plan that she thinks drives this loosely organized global movement she calls "the shock doctrine."[90]

Klein's thesis is overextended because she too frequently confuses malign results with malign intent. But her work deserves our attention here because in some circumstances, particularly those involving coastal villages after the Asian Tsunami, she is right. In other circumstances, such as New Orleans after Katrina, her observations show how quickly even some well-intentioned initiatives can descend into government ineptitude and social injustice.

Consider Arugam Bay, located on Sri Lanka's southeast coast. On that sandy crescent you could once find a thriving fishing village in addition to some of the best surfing on the Indian subcontinent. These uses did not always coincide. Local hotel owners considered the fisher folk something of a bother. Their grass huts blocked ocean views. The smell of drying fish drove sensitive guests away. After the 2002 cease-fire between the Sinhalese government and the Tamil Tigers, hoteliers and developers dreamed of transforming this faded fishing port into a sparkling tourist destination. But even with the national government's support, there was no politically acceptable way to do it. The hundreds of fishing families on Arugam Bay had lived there for generations. They had nowhere else to go.

It is said the villagers "mingled easily" with foreign tourists, mostly young Australians in board shorts and puka beads. But the community showed grave distrust for developers, and the Sri Lankan government too. This was, at best, an uneasy détente. But the 2004 tsunami changed all that, destroying most of the village and leaving more than 2 million people in the region without homes. Before the tsunami, there appeared to be no politically viable way to do it. Now the resort developers had what they had wanted all along: "a pristine beach (in a prime area), scrubbed clean of all the messy signs of people working, a vacation Eden."[91]

Sensing an opening, the government moved swiftly. It passed new public safety laws and created shoreline "buffer zones" in which the rebuilding of traditional housing was prohibited. Homeless Sri Lankans were moved inland to metal-roofed barracks, patrolled by "menacing machine-gun wielding soldiers." Resort developers were treated differently. They were given incentives to expand their operations onto the newly vacated oceanfront land. A new "Task Force to Rebuild" was established that included many of Sri Lanka's most powerful banking and business leaders and excluded villagers, farmers, and environmentalists. Government officials began dusting off development plans that had been in the works for years for the transformation of Sri Lanka's war-torn coast into a series of tourist zones—a "Bamboo Riviera"—with five-star hotels, luxury chalets, and world-class shopping. Sri Lanka's president, Chandrika Kumaratunga, who before the tsunami had been elected on an "antiprivatization platform," suddenly began privatizing water works and electric plants. The U.S. government backed this redevelopment plan and was soon sending in private companies like CH2M Hill to build bridges and dredge harbors in preparation for the coast's new resort economy. By helping to grow the economy, American aid officials believed such development would help all Sri Lankans, rich and poor alike. But Klein reports that the displaced villagers did not always see it that way, particularly when forced to turn over the land they had lived on to Hilton or Marriott. According to Klein, the only direct American financial aid that small-scale fishers ever received was "a $1 million grant to 'upgrade' the temporary shelters where they were being warehoused while the beaches were being developed."[92]

The Sri Lankan experience appears to be the rule rather than the exception in countries like India, Thailand, Indonesia, and the Maldives. And it is not just Klein who says so. An ActionAid survey of fishing villages after the tsunami in all these countries and in Sri Lanka documents similar stories of land seizures, forced relocations, and lavish incentives for resort development. As in Sri Lanka, aid money was used to fuel much of the development, leaving many of the tsunami's neediest victims empty-handed and footing the bill. And as in Sri Lanka, the result was not a matter of simple corruption or miscalculation but of intentional planning. The report concludes: "Governments have largely failed in their responsibility to provide land for permanent housing. They have stood by or been complicit as land has been grabbed and coastal communities pushed aside in favour of commercial interests."[93]

Back in New Orleans, the adventure in disaster capitalism worked this way. As Hurricane Katrina approached the Gulf Coast, most middle- and upper-class residents fled the city by car, checked into hotels, and watched the storm's aftermath on CNN. People in this group also had the resources to return quickly to the city once water and electricity were again running. If their homes were located in areas that did not flood, they were able to resume their lives with at least some degree of normalcy. For others, the experience was more uncertain. Those left stranded in the Superdome and Convention Center were later transported out of the city and scattered in cities across the country, from Atlanta to Las Vegas. Other poorer residents, lucky enough to have evacuated on their own, would discover their homes had been destroyed and would lack the money or insurance coverage to immediately return and rebuild.

It was against this backdrop that the aging Milton Friedman informed readers on the opinions page of the *Wall Street Journal* that he saw in the Katrina disaster "an opportunity."[94] Weeks later, a group of free-market economists organized by the conservative Heritage Foundation produced a list of "Pro-free-market Ideas for Responding to Hurricane Katrina and High Gas Prices." This list, which they called a "hurricane relief" package, called for suspending the prevailing wage laws, creating flat-tax "free-enterprise" zones, waiving the estate tax (called a "death tax") for Katrina victims, and issuing school choice vouchers for displaced children. These economists also called for waiving environmental assessments, suspending environmental regulations, permanently cutting the gasoline tax, and drilling in the Arctic National Wildlife Refuge. Some of these ideas, including the flat-tax zones, the suspension of wage laws, and the suspension of air pollution standards, were eventually adopted.[95]

It was as if a mighty hand had stretched down from the clouds and pushed the reset button for the entire Gulf Coast. Lucrative contracts for cleanup and rebuilding were awarded to firms like Halliburton, Bechtel, and Blackwater USA, in a noncompetitive process. In contrast to the largesse received by Halliburton and other firms, the Bush administration refused emergency funds for public sector salaries, a move that resulted in the firing of more than 3,000 city employees. Three historic public housing structures that federal officials had long wanted to close were boarded up and eventually demolished. In describing these events, Representative Richard Baker, a 10-term Republican from Baton Rouge, told a group of lobbyists, "We finally cleaned up public housing in New Orleans. We couldn't do it, but God did."[96]

Just as in the aftermath of San Francisco's 1906 earthquake, some city leaders began thinking of ways take advantage of Katrina as an opportunity to redesign the city for the benefit of developers and entrepreneurs. In San Francisco, the city's elders hoped to spur development by relocating Chinatown. In New Orleans, bars and coffeehouses were abuzz with stories of city elites plotting to rid the city of its "undesirable elements," by which one could mean criminals, the unemployed, African Americans, or Democrats, depending on the speaker.[97] Not everyone, of course, was thrilled—certainly not the tens of thousands of residents wondering how they would be able to return and rebuild. Reverend Jesse Jackson famously called Katrina "a hurricane for the poor and a windfall for the rich."[98]

These examples do not always add up to conspiracy. Arugam Bay might. There, the case for collusion between the Sri Lankan government and the international resort industry seems particularly convincing. But even though the Katrina disaster may have struck some mad visionaries as a grand opportunity to be tweaked and manipulated, many of the government's omissions and delays seem embarrassingly unplanned. For me the government's misadventures in New Orleans serve more as examples of cronyism, carelessness, and incompetence. Moreover, the "shock doctrine" thesis does not credit the enormous good that private investment can achieve if channeled effectively.

But Klein is correct to draw a connection between market capitalism and the burdens borne by the poor in times crisis. Disasters clear the board. Whatever influences or mindsets that exist outside the boundaries of the resulting destruction will move in. If that mindset is unrestrained capitalism, the new order will bear resemblance to Klein's disaster capitalism. If that mindset is a commitment to fairness and the social contract, the new order will reflect concern for vulnerable people and will seek to counterbalance the worst impulses of the private market.

Bias and Bigotry

Justice, of course, is not just in tension with the material values of the market; it is also at odds with more visceral emotions of hate and resentment, qualities we identify with racism and other forms of bigotry. At the height of government rescue efforts after the flood, President George W. Bush seemed not to have contemplated the importance of social vulnerability in times of crisis. When a reporter asked him about the slow pace of the emergency response and its disastrous effect on African Americans, he switched the focus from the *effect* of the slow recovery to the *intent* of the responders. "When those Coast Guard choppers . . . were pulling people off roofs," he huffed, "they didn't check the color of a person's skin. They wanted to save lives."[99] Of course, no one was arguing that the Coast Guard had intentionally discriminated against anyone. (Indeed, the Guard's performance was exemplary.) Rather, the reporter was calling attention to a particular problem associated with slow response time: because blacks were more exposed and more vulnerable, the *unintentional* effect of a slow response was that the resulting harm was borne disproportionately by people of color. Speaking at a remembrance service four days later at Washington National Cathedral, Bush appeared to have grasped the point. "The greatest hardship fell upon citizens already facing lives of struggle," he said mournfully; "the elderly, the vulnerable and the poor. And this poverty has roots in generations of segregation and discrimination that closed many doors of opportunity."[100]

Bigotry—in the form of racism, sexism, or anything else—is hard to talk about in America because definitions are so hard to pin down. Does a racist motive *require* its holder to feel fear or hate? Can *institutions* be racist if their human constituents are not? And most perplexing, can people *act* on racist motives without *knowing* it? I can't answer these questions, but I can organize a framework that helps us understand the issues involved. We should start by acknowledging that there is a range of objectionable behavior associated with what is commonly called discrimination. Some of that behavior follows from a conscious and invidious bias (*intentional* bias). Sometimes decisions that disadvantage a particular group are unintentional but rooted in a structure once influenced by intentional bias (*structural* bias). Finally, decision-makers can act on irrational stereotypes or bias without being aware that they are doing so. This form of unconscious behavior is at the same time unintentional *and* motivated by bias (*unconscious* bias). While all of these forms deserve our attention, I believe the law's condemnation of discrimination should be related not to the degree of motive but to the moral or social harm that it causes.

INTENTIONAL BIAS

The most universally condemned form of bigotry seems to be that which is specifically intended, sometimes out of fear or hate. This form of discrimination often rears up in cases of disaster and in the recovery that follows. Such vileness inspired white officers to detain black laborers at gunpoint during the Great Mississippi Flood of 1927 and whipped up violent assaults against the Japanese in postquake San Francisco (Chapter 5). After Katrina, there were many charges of intentional racism leveled against state and private actors. The Gretna bridge standoff was one event that inspired such charges. As noted, legal complaints against Gretna officials, some of which involve allegations of intentional racism, have yet to be resolved. Months after the flood, in neighboring St. Bernard Parish (a jurisdiction that was 93 percent white before Katrina), lawmakers were accused of racism when they passed an ordinance requiring council permission for any owner to rent a single-family home to anyone not a blood relative. A housing advocacy group sued the parish under the Fair Housing Act; the case was eventually settled and the ordinance revised.[101]

Many people believe that intentional racism played a decisive roles on Gretna bridge and in St. Bernard's parish council; but that conclusion is difficult to prove in court. That is the problem: even if it seems very likely that a police officer or a parish council member was motivated by racial bias, a system that requires evidence of such motivation to establish relief will leave too many instances of unfairness unaddressed. No doubt the reason St. Bernard Parish settled the case and rewrote the ordinance was that the Fair Housing Act did not require the plaintiff to establish a discriminatory *intent* but only a potentially discriminatory *outcome*. When statutes like these do not exist, as in the environmental justice struggles involving the placement of landfills in minority neighborhoods (Chapter 5), plaintiffs are left with having to prove an intended racial bias, and they lose.

We should also note that not all instances of intentional racism should be regarded as equally harmful or dangerous. When Nagin (who is black) promised a crowd five months after the storm that "[t]his city will be chocolate at the end of the day. This city will be a majority African-American city. It's the way God wants it to be," there was little question that his comment was both racist and ridiculous.[102] But there was also little evidence that the mayor, who often enjoyed the backing of white businesspeople, was seriously harming anyone other than himself. What all of this means is that while motive is relevant to evaluating the harm of discrimination, it does not *by itself* make a discriminatory act more alarming or important than some other race-based harm for which a bad motive cannot be shown. And even where the presence of intentional bias (as perhaps existed in St. Bernard Parish) does elicit one's disgust, it may not be practical to focus on that element as part of the legal means to combat it.

STRUCTURAL BIAS

In democracies, the most destructive part of bigotry in disasters is the structural kind. Systemic policies and patterns of distribution exist that affect certain groups as defined by race, sex, or other characteristics. These might include the segregated residential patterns, disparities in education, and economic policies that widen the gap between rich and poor and make it increasingly hard to leap across that canyon. In the case of sex, institutional bias can be seen in an educational system that alienates girls from math and science, employment policies that burden primary caretakers, and property laws that have the effect of enriching the material lives of men while doing the reverse for women.[103] Law professor Richard Thompson Ford, who has examined the influence of geographic segregation and voting patterns on race, calls this "racism without racists."[104] We could just as easily call this bigotry without personal bias.

In diagnosing the racial injustices of Katrina, Ford attributes most of the problem to the factors I noted earlier in the chapter—the concentration of African Americans in flood-prone parts of the city, the relative poverty of African-American households, and their low rate of car ownership. All of these problems, Ford notes, are rooted in the history of unequal treatment that is now perpetuated in the very structure of our society. Generations of restrictive land-use laws, discriminatory lending patterns, and white intimidation have shunted blacks into undesirable residential enclaves. After the original racist motives weakened (or went underground) social and market forces continued to reinforce the pattern. In the same way, segregation in education and employment has forced blacks into a second-class economy that has perpetuated itself generations beyond the official banning of such practices. Ford says of the political leadership at the time of Katrina: "[i]t's natural to want to hold the available blameworthy parties responsible for *all* of those evils. But most of the racists responsible for the distinctly *racial* cast of the Katrina disaster are dead and gone."[105]

Ford is correct, but only to a point. What he says about structural racism is true. Generations of racial bias have indeed put our housing and employment markets on a kind of race-based "autopilot" that even a complete eradication of conscious racism could not now destabilize. For that reason, policy-makers seeking racial equality must look beyond punishing racial motive and instead focus on progressive economic and regulatory reforms aimed at actively resetting the balance. But Ford goes too far to suggest that "most of the racists responsible" are "dead and gone." That is not the lesson I take from the Gretna bridge standoff or from lawmakers' attempt in St. Bernard Parish to exlude people "not related by blood."

Moreover, the "dead and gone" thesis dilutes the objection we should voice against what I might call "passive bias": habitual indifference to a vulnerable group based in part on members' race, sex, or other attribute. Here we must turn to rapper Kanye West's famous assessment of President Bush on the subject of Hurricane Katrina. Five days after the levees broke, West and comic actor Michael Myers hosted a benefit concert that was broadcast on live television to help victims of Katrina. At one point the two stood stiffly on stage as video footage of rescue helicopters hovered silently behind them. Reading carefully from a teleprompter, Myers intoned, "The landscape of the city has changed dramatically, tragically, and perhaps irreversibly. There is now over twenty feet of water where there was once city streets and thriving neighborhoods." Then it was West's turn. "I hate the way they portray us in the media," he said, ignoring the script. "You see a black family, it says, 'They're looting.' You see a white family, it says, 'They're looking for food.' And, you know, it's been five days [waiting for federal help] because most of the people are black." Myers, obviously stunned, stared into the camera and continued his lines. "The destruction of the spirit of the people of southern Louisiana and Mississippi may end up being the most tragic loss of all." West then said, spontaneously, "George Bush doesn't care about black people!"[106]

Some people, including Richard Ford, took that to mean that West was calling the president a racist.[107] In fairness, the statement that Bush "doesn't care about black people" might be read to suggest that his conduct betrayed not an active bias against blacks but simply a passive indifference toward them. Jacob Weisberg, an editor of *Slate*, made this point at the time: "Because they don't see blacks as a . . . constituency, Bush and his fellow Republicans do not respond out of the instinct of self-interest when dealing with their concerns. . . . Had the residents of New Orleans been white Republicans in a state that mattered politically, instead of poor blacks in a city that didn't, Bush's response surely would have been different."[108] Weisberg's self-interest argument makes intuitive sense. In the 2004 presidential election, Bush's share of the black vote barely cracked 10 percent.[109] In addition to explaining the Bush administration's indifference to the needs of African Americans, it might also explain its indifference and even hostility toward urban needs like air regulation and public infrastructure, which I described earlier in this chapter.[110] In fact, Weisberg's argument resembles law professor Derrick Bell's "interest-convergence" thesis—a staple of Critical Race Theory—which holds that racial majorities will support racial justice only to the extent that there is something in it for them.[111]

Assuming that politicians respond in such calculated ways, what remains for us to explain is why Bush apparently saw no chance for political glory in riding to the rescue of what must be described as one of the most needy and vulnerable populations in urban America. The answer must be that he didn't think most Americans cared about black people either. Perhaps he was right. For while the Katrina debacle undeniably contributed to the public's dissatisfaction with Bush, that displeasure appeared to reflect more a concern for general competence than for Katrina's victims. Otherwise, one would have expected a demand for more recovery aid and more media attention.

If the president's passive bias toward blacks (or other groups) was, in fact, a reflection of Americans' own passive bias toward blacks (or other groups), then the problem goes far beyond Katrina and the Bush presidency. The next time a major flood or earthquake strikes and vulnerable communities are again ravaged, the same collective indifference that allowed Bush to avoid bold action would similarly let other leaders off the hook. This would be true even for leaders who have great sympathy for the disadvantaged. Without strong public support, it is hard for any leader to muster the resources to address disaster injustices on the ground.

UNCONSCIOUS BIAS

Is it true that Americans show less concern for some racial groups in times of disaster? And if so, can anything be done about it? Shanto Iyengar, a communications professor at Stanford University, and Richard Morin, a polling expert at the *Washington Post*, started asking questions like these in the months following Katrina. Specifically, they wanted to know if racial cues conveyed in news coverage had influenced Americans' response to government relief efforts. Was it possible, they wondered, that public outrage over the national response might have been softened by knowledge that many victims of the flood were black? To learn more, they designed an online experiment conducted through the web site of the *Washington Post*.[112] As part of the experiment, online participants were asked to read a news report describing the impact of Hurricane Katrina and then complete a brief opinion survey. More than 2,000 readers completed the study. In terms of demographic characteristics such as age, gender, and ethnicity, the participants pretty much reflected the national population. But as regular fans of the *Post*, they were more educated (84 percent had bachelor's degrees) and more likely to identify themselves as Democrats or liberals; 86 percent of respondents were also critical of Bush's handling of the Katrina disaster.

Unknown to participants, the news report was fictional. It told the typical story of a displaced Katrina victim who hoped to resume a normal life. The sex and ethnicity of the victim were varied in the stories to see if such details would elicit different responses among readers. Each news report also contained a photograph of the victim, either male or female. The photographs were edited through a process of "digital blending" to vary the subject's ethnicity and skin complexion. The featured subject could thus be a man or a woman who was either white, African American, Hispanic, or Asian. In any category, the subject could be of "light" or "dark" skin tone. The survey, given after the report had been read, asked participants a series of questions about the level of public or private assistance they thought the victim should receive.

Iyengar and Morin found that even with a cohort sharply skewed toward liberals and college graduates, race and skin tone affected the outcome in significant ways. The survey, for instance, asked respondents whether assistance to hurricane victims should be mostly borne by the federal government or by private charities. "Respondents who encountered an African-American victim were on average roughly 6 percent less likely to nominate the federal government and 3 percent more likely to suggest charities and individual victims as the major source of disaster relief when compared with those who encountered a white victim." These results, the reviewers concluded, "suggest that public support for large-scale governmental relief efforts is weakened when hurricane victims are disproportionately African American." Answers also varied on the amount and longevity of assistance. Respondents who viewed the white victims generally favored more generous assistance packages than respondents who had viewed nonwhite victims. In addition, within nonwhite racial groups, the level of aid generally declined for victims whose skin tone was darker; light-skinned blacks or Hispanics elicited more generous assistance packages than their dark-skinned counterparts. (In contrast, whites induced more generosity when they were well tanned.) "The fact that this group awarded lower levels of hurricane assistance after . . . encountering an African-American family displaced by the hurricane is testimony to the persistent and primordial power of racial imagery in American life." Iyengar and Morin do not speculate as to whether the racial preferences were consciously or unconsciously held. Perhaps both kinds of bias played a role. But because the education level and liberal leanings of the survey group are traits that tend to correspond with a lower level of self-identified prejudice, it seems likely that a more subliminal form of bias was also skewing the results.

Critical race theorists have emphasized the role of subliminal bias, sometimes borrowing from Sigmund Freud and his theories of the unconscious.[113] Some environmental justice scholars have also picked up on the idea, suggesting that unconscious prejudice might account for planning decisions that put environmental hazards in minority neighborhoods.[114] But legal scholars have only recently begun tying these theories to the very rich work of the social psychologists who have been studying unconscious prejudice since the 1970s.[115]

The study of unconscious bias is seeping into popular culture, and vice versa. Millions of people have taken the computerized "Implicit Association Test," offered by Project Implicit, a long-term research project developed by psychology professor Mahzarin Banaji and based at Harvard University.[116] The test measures respondents' accuracy and reaction times when told to associate words like "joy" or "awful" with the faces of either blacks or whites. If a person takes longer to associate positive words with blacks than with whites, that is thought to imply an implicit bias toward whites. Seven years of operation show that "75–80 percent of self-identified Whites and Asians show an implicit preference for racial White relative to Black."[117] (Interestingly, about half of all black respondents also show an implicit preference for whites.) The 2008 presidential election led to many studies about the relationship between race and national identity. A study conducted by social psychologists Thierry Devos and Debbie Ma before the 2008 presidential election found that respondents (all California college students) unconsciously perceived brown-skinned Barack Obama to be less American than the pink-skinned Tony Blair.[118] A similar study by Devos and Ma showed that although people know that actor Lucy Liu is American and that actor Kate Winslet is British, respondents' minds unconsciously associated Liu's Asian features with being foreign and Winslet's Caucasian features with being American.[119]

Social psychologists continue to debate the reliability of surveys like these, as well as the precise conclusions that can be drawn from them. But legal theorists who study unconscious bias believe that at the very least, this research shows a basic failing in the law's understanding of identity-based discrimination. Linda Hamilton Krieger, a law professor at Berkeley, is a leader in the field. She points out that antidiscrimination law generally proceeds from the premise that discrimination on the basis of race, sex, or any other factor is consciously employed.[120] Thus, to establish illegal discrimination under the equal protection clause of the Fourteenth Amendment, a plaintiff must show that the defendant acted with discriminatory intent.[121] Because it is often difficult to establish such intent, even where it exists, Congress has tinkered with the burden of persuasion in antidiscrimination statutes involving employment, housing, and education. Under such laws, a plaintiff may establish a prima facie case of discrimination by proving a discriminatory outcome; the burden then shifts, and the defendant must prove that he has some legitimate reason for treating a person differently that is not simply a pretext for unlawful discrimination. But even this presumes what Krieger calls "transparency of mind," that is to say, a conscious awareness of the motives one is actually acting on.[122] This presumption is very important in antidiscrimination law because it suggests that the way for society to discourage discriminatory behavior is to identify those bad actors who (consciously) discriminate and punish them by holding them responsible for the damage they cause. The idea is that once everyone believes that the law will find bad actors and hold them accountable, would-be discriminators will stop their bad behavior and act like good citizens.[123] But if, as social cognition theory suggests, decision-makers are unaware of their motivating biases, this project is doomed to failure. A FEMA official might still lowball a minority community's compensation package or a rescuer might still avoid the barrio without even being aware that a bias is being acted on.

The situation, however, is not hopeless. The research on cognitive behavior suggests that people can *correct* for unconscious bias. This finding lies at the heart of the recent legal scholarship on the subject. As Krieger, again, explains: "Cognitive biases in intergroup perception and judgment, though unintentional and largely unconscious, can be recognized and prevented by a decisionmaker who is motivated not to discriminate and who is provided with the tools required to translate that motivation into action."[124] Basically, these tools are institutional practices that require decision-makers to think carefully about their motivations ahead of time and to consider the perspectives of groups with whom they are less familiar. This is why the designers of the "Implicit Association Test" require you to make word-and-face associations as quickly as you can without thinking: if you were allowed to slow down and give each response careful thought, you would move from an unconscious to a conscious mode of reasoning, thus keeping your latent stereotypes in check. The strategy resembles the "stop and think" model used in the National Environmental Policy Act, in which federal decision-makers are required to consider environmental impact when reviewing major projects.

To better avoid discrimination in the context of disaster policy, implicit bias theory suggests that the law should move beyond strategies designed to smoke out consciously hidden bias and instead look for ways to prevent unconscious bias from hijacking the system. The solutions need not be fancy. They may simply require decision-makers to consider more carefully the reasons behind the choices they make and to make sure they consider issues of particular importance to outside groups. I will develop this strategy further when I propose in Chapter 7 an executive order on disaster justice.

7

WINDS OF CHANGE

How can law help pursue disaster justice? A disaster justice approach in public policy would borrow the insights we saw in the environmental justice movement. It would build an affirmative agenda but also make use of existing law and precedent to urge government accountability. Most prominently, it would be based on the interconnections between environmental exposure and social vulnerability.

This might begin with the way academic research is conducted. As a report by the Sage Foundation noted, environmental sociology and disaster research are often developed separately. One doesn't learn much about acute, fast-moving catastrophes in environmental sociology books, nor does the disaster literature refer very much to environmental justice research. That is a shame, since the goals of both disciplines are the same: "to use systematic and thorough research to uncover inequality in exposure to hazards and risks, and to support organizing and policy change to reduce risk and suffering."[1] When studying disasters, scholars should pay attention not only to the hazardous agents (the quake, the fire) and levels of exposure but also other factors like neighborhood contamination, preexisting disabilities, poverty, and bigotry.

This awareness could lead to changes in disaster policies that integrate and emphasize the role of social inequality and financial need. Disaster policy must also develop a fuller affirmative agenda. Environmental justice advocates have, through their own hard work, gradually learned to build up their movement's affirmative agenda. Environmental justice is now about more than reactive justice on the microlevel—opposing expansion of the local landfill or a broader emissions permit for the neighborhood refinery. It also encompasses "proactive approaches aimed at transforming the underlying structural causes of environmental inequality, economic inequities, and health disparities."[2] Disaster justice should do likewise, advocating a change from so-called hazard-by-hazard planning to a holistic strategy that incorporates cumulative and synergistic exposures and vulnerabilities.

Over time, environmental justice advocates have developed a "toolbox" of legal resources that they use to support the social and political side of their struggles. We need a disaster justice toolbox, too, that borrows some of the laws the environmental justice movement has used successfully and adds new laws and strategies to suit more specialized needs. Tools should be drawn from both domestic law in the United States and international law, for use in the United States or in other countries. On the domestic side, a disaster justice toolbox would draw from two sources: (1) laws protecting health, welfare, and the environment, and (2) antidiscrimination law. After examining the domestic side, I'll move on to international law.

Domestic Law

LAWS PROTECTING HEALTH, WELFARE, AND THE ENVIRONMENT

Following the adage that "a strong community is a healthy community," the first prescription should be to shore up the traditional laws that protect health, welfare, and the environment. A society's ability to endure and re-cover from shock will ultimately depend on the long-term policies it has put in place to promote a healthy citizenry, adaptive economies, and produc-tive ecosystems.

Perhaps the most difficult prescriptions entail broad improvements in the programs that generally help the disadvantaged. Improved job security, more generous unemployment insurance, universal health care, and better pub-lic housing might sound like "pie in the sky," but these are the *real* ways to lower community susceptibility and increase resilience.[3] It's worth recalling that Lyndon Johnson's investments in poverty reduction and other aspects of "social infrastructure" paid large dividends, reducing the portion of Americans living in poverty from 21 to 12 percent in about seven years.[4] The generations of neoliberalism that followed never did match that record (some would say the point was not to try) and succeeded only in widening the country's wealth gap and increasing the vulnerability of America's mar-ginalized populations. In addition, the policy of letting states provide differ-ent levels of federal unemployment benefits (independent of the regional cost-of-living standards) should be abandoned. It is simply unacceptable that states like Louisiana, Mississippi, and Florida—whose hurricane-prone shores house a large share of the nation's poor and minority populations—

should also have the lowest unemployment benefits in the country. For this reason the Obama administration deserves credit for using state stimulus money as an incentive for states to raise their unemployment benefits.

Similarly, the government must renew its commitment to strong health and safety regulations. Over the years, many Americans have lost confidence in the basic safety of medicines, food, and consumer products. In 2004 it was revealed that at least 88,000 people suffered strokes or heart attacks induced by an antiinflammatory drug Vioxx. Experts proposed many explanations for how this happened, but one prominent problem appears to be the Food and Drug Administration's "inability [because of a lack of resources and leadership] to monitor the safety of drugs after they are on the market."[5] More recently, the agency came under fire for failing to take prompt action against unsafe peanut processors, a failure that resulted in an outbreak of salmonella in peanut butter that sickened more than 500 people in the United States and Canada and killed at least eight. In response, President Obama promised "a complete review of FDA operations," noting that it "has not been able to catch some of these things as quickly as I expect them to catch [them]."[6] In 2007, discovery of toys containing lead and other unsafe products imported from China led to a rash of recalls and lawsuits. A primary contributing cause appeared to be vast underresourcing of the Consumer Products Safety Commission, which lacked sufficient staff "to monitor the safety of the growing flood of imports."[7] Reports of regulatory agencies failing to protect the public because of scant resources or poor leadership have become common and require sustained remedial attention.

In addition to these general prescriptions, disaster justice advocates should pay special attention to environmental protection in areas prone to natural hazard. They should do so for two reasons. First, all things being equal, there is a greater chance in such areas that the public will be exposed to improperly stored or treated contaminants because of the danger of storms, earthquakes, or the like causing ruptures. Second, people living in areas with bad environmental protection may already be faced with compromised health, higher stress, or devalued land, and these hardships make them more vulnerable to natural hazards when they occur.

In particular, regulators and lawmakers should focus on toxins. New Orleans and Galveston were both inundated with toxic spills when they were flooded by storm surge. These spills came from active facilities like such as that of Murphy Oil in St. Bernard Parish, which are regulated under the federal Resource Conservation and Recovery Act. Had enforcement of storage and treatment rules been tighter, some of these spills might have been avoided.[8] In one positive development, the EPA, pursuant to the Executive Order on Environmental Justice and Title VI of the Civil Rights Act, now includes environmental justice as a factor in reviewing state permits issued under the Resource Conservation and Recovery Act as a means of protecting vulnerable communities from bearing an unfair share of toxic facilities. (The EPA is doing the same with air emission permits issued under the Clean Air Act.) But some vulnerable chemicals are not even regulated under federal law. In December 2008, after 10 days of pounding rain, an earthen dike burst at a coal-fired power plant 40 miles outside of Knoxville, Tennessee. More than 5 million cubic yards of wet coal ash swept over thousands of acres, including a residential neighborhood. While the Tennessee Valley Authority, which operated the plant, insisted that the local drinking water was not at risk, the EPA warned that the sludge might contain arsenic, lead, and other toxic metals. A test of river water near the spill showed elevated levels of lead and thallium, which can cause birth defects and nervous and reproductive system disorders. Perhaps the saddest news was that the spill would never have happened if coal ash were regulated under the Resource Conservation and Recovery Act. Rather than storing its wet ash in special lined landfills, the plant was instead allowed to dump its waste in unlined embankments and open ponds.[9]

Many environmental justice advocates have also focused on reporting requirements as a way to discourage more toxic emissions in poor and minority neighborhoods. For instance, Massachusetts's Toxics Use Reduction Act requires facilities to inventory chemicals entering and exiting their production cycles and to develop a "toxics use reduction plan." While the law does not require facilities to implement the plan, the exercise and the public availability of the information has led firms to reduce their use of toxic chemicals significantly.[10] At a national level, the Toxics Release Inventory—a program created in the aftermath of the 1986 chemical plant explosion in Bhopal, India—requires large polluters to report annually their releases into the air, ground, and water. Despite some weaknesses, the inventory has helped educate and mobilize the public and, as a result, caused polluters to reduce emissions. The Bush administration had sought to weaken reporting requirements out of concerns for cost and protection against terrorism. But in the Obama administration, this program will—one hopes—be maintained and expanded.

Another source of dangerous toxins is abandoned sites. Recall that in the aftermath of Hurricane Katrina, floodwaters inundated at least three Superfund sites located near poor African-American communities in New Orleans (Chapter 7). In 2008, Hurricane Ike swept across 29 Superfund sites in Louisiana and Texas. Almost immediately, officials began worrying about dioxin running from the San Jacinto Waste Pits site into Galveston Bay, threatening shoreline communities and local fisheries.[11] Superfund sites are, by definition, the most seriously contaminated facilities in the country. Once identified and listed, the Superfund law requires that such a facility be cleaned up. If parties related to the contaminated site cannot be found or are insolvent, monies for the cleanups are to be taken from a special trust, the "Superfund," funded by a tax on the chemical industry. That tax was allowed to lapse in 1995. The Superfund is now depleted, and cleanup of "orphaned" sites depends on an unsteady stream of appropriations from the general federal revenue.[12] As a result, the rate of Superfund cleanups has fallen abruptly. From 1992 to 2000, for instance, the number of completed cleanups averaged around 77. From 2001 to 2005, the average was 42.[13]

This slow rate of Superfund cleanups threatens public health across the country. One in four Americans lives within three miles of a Superfund site. Roughly 3–4 million children, whose bodies are more vulnerable to many environmental poisons, live within 1 mile. To call attention to the problem, the Center for Progressive Reform reviewed the five "worst" Superfund sites in each of the 10 most populous states. Data from these 50 sites confirm many of the things health and environmental experts would expect. Most of the sites, for instance, are in heavily populated urban or suburban neighborhoods. People of color and those of lower income are disproportionately located around a significant number of these sites. Large numbers of children and elderly people also live near these facilities, making a total of "34,127 children aged 9 and younger and 14,068 persons aged 75 and older."[14]

A review of the report also suggests that many of the identified sites are also in regions threatened by serious natural hazards. For instance, at the American Creosote Works near Pensacola, Florida, a former wood-preserving facility contaminated with creosote and pentachlorophenol, ponds set up to "percolate" these highly toxic liquids have a history of overflowing and spilling into Bayou Chico and the Pensacola Bay. The surrounding census tract is 48 percent minority and has a median household income of $23,000. The

Pensacola area is one of the most active hurricane regions in the country. When Hurricane Ivan blasted ashore on September 16, 2004, Pensacola was the city hardest hit. On average, the Pensacola area experiences a hurricane about once every eight years and is "brushed" or hit by a tropical storm on average once every three years.[15]

In Gynn County, Georgia, the 550-acre LCP Chemical site was used for 70 years as an oil refinery, paint manufacturing plant, power plant, and chloralkali factory. The EPA estimates that more than 380,000 pounds of highly toxic mercury was "lost" in the area between 1955 and 1979; as a result, commercial fishing in the area is banned. The census tract is 63 percent minority and has a median household income of $24,000. Like Pensacola, Glynn County is hurricane country and has been debilitated by many storms in its history, most recently Hurricane Ivan (2004) and Hurricane Francis (2004).[16]

The same report lists other Superfund sites that should be of concern to disaster justice advocates, including the former Stringfellow disposal site in Glen Avon, California (an area prone to earthquakes), the McCormick & Baxter Creosote Company site near California's San Joaquin River (earthquakes, floods, and levee ruptures), the Lawrence Aviation Industries site in Suffolk County, New York (hurricanes), and the Normandy Parks Apartments site in Hillsborough County, Florida (hurricanes).[17]

These findings suggest the need for a more aggressive approach to cleaning up Superfund sites, particularly those located in areas where natural hazards increase the likelihood of human exposure. The obvious first step is for Congress to reinstate the Superfund revenue stream, which once generated tens of millions of dollars for cleanup efforts. Indeed, after Hurricane Ike raised questions about dioxin contamination in the San Jacinto River, Congressman Gene Green of Houston called for this action.[18] Next, Congress and the EPA must ensure that the resulting cleanups are not only prompt but also adequate to protect against some of the worst catastrophes.

ANTIDISCRIMINATION LAW

In Chapter 5, I described the role antidiscrimination law has played in the environmental justice movement. As a tool, antidiscrimination law has proved occasionally helpful to environmental justice but has failed to live up to advocates' original expectations. For similar reasons, federal antidiscrimination law as it is now interpreted offers only marginal hope to disaster justice advocates.

To review the discussion in Chapter 5, we've seen that some antidiscrimination laws, like the Fourteenth Amendment's equal protection clause, bar only *intentional* discrimination. The statutory alternative most relevant to environmental justice, Title VI, has been judicially interpreted to allow bars on both intentional *and* unintentional discrimination (that is, disproportionate outcomes based on race) but provides no private remedy for the latter. (So far, no court has permitted a private remedy for unintentional discrimination through Section 1983 either.) Thus environmental justice advocates confront the painful dilemma that the law that recognizes unintentional discrimination gives them no right to sue, while the law that lets them sue gives them no protection against unintentional discrimination. Although *some* antidiscrimination statutes, like those governing education, employment, and housing, bar unintentional discrimination *and* provide private remedies, such laws are of limited use to people whose grievances are based on environmental harm.

Victims of discrimination related to disaster response and recovery face the same problem. Suppose a group of plaintiffs believes that federal rescuers spent more resources in suburban neighborhoods than urban neighborhoods and that that choice produced a racially discriminatory outcome. Under the equal protection clause, the plaintiffs would have to show that government decision-makers were motivated by race. Under Section 602 of Title VI, plaintiffs would have no private right of action at all. Indeed, the situation is even bleaker. Unlike the implementing regulations followed by EPA in interpreting Title VI, FEMA's implementing regulations do not even make clear that it believes that Title VII even applies to unintentional discrimination.[19] Some claimants in the disaster context might be more able to fit their grievances into the categories of education, employment, or housing, thus taking advantage of more favorable laws. But for those who depend on the more general Title VI, the prospects are slim.

Faced with this reality, disaster justice advocates might look to another law more directly related to emergency response: the Robert T. Stafford Disaster Relief and Emergency Assistance Act of 1988, which governs the way federal resources can be used in responding to major disasters or emergencies. The act provides both antidiscrimination measures and substantive rules that disaster justice advocates could use. Section 308 of the Act, for instance, requires regulations of implementation to "include provisions for insuring that the distribution of supplies, the processing of applications, and other relief and assistance activities shall be accomplished in an equitable and impartial manner, without discrimination on the grounds of race, color, religion, nationality, sex, age, disability, English proficiency, or economic status."[20] Section 308 further requires that any governmental body or private organization participating in the federal relief effort must comply with these nondiscrimination rules.[21]

Section 308 should be praised for expanding the categories used to define discrimination and for including private organizations within its scope. But nothing in the text of the law addresses or suggests whether Section 308 would cover unintentional discrimination. At the time of this writing, FEMA has not promulgated regulations to implement Section 308, so there is no way to know how the agency will interpret it. When FEMA does propose its regulations, it should make clear that it reads the law to bar both intentional and unintentional discrimination. If it does not, the burgeoning disaster justice community should urge reform in the administrative notice-and-comment process. But even if agency regulations read Section 308 to apply to unintentional discrimination, the section's text still does not explicitly provide a private remedy. Moreover, it seems very unlikely that a court would infer one, since there is little evidence in either the section's phrasing or its legislative history that suggests that Congress intended to create a private remedy.[22]

Although the Stafford Act contains no private remedy for its antidiscrimination provision, it does offer other features of interest to disaster justice advocates. Section 512, for instance, explicitly requires state and local governments to develop evacuation plans that take vulnerable populations into account. Policy-makers must coordinate and integrate evacuation plans "for all populations including for those individuals located in hospitals, nursing homes, and other institutional living facilities."[23] Procedures must be developed for communicating these plans to members of the public, including those "with disabilities or other special needs," those "with limited English proficiency," or those "who might otherwise have difficulty in obtaining such information."[24] By requiring substantive actions, this section prompts planners to take into consideration factors that even well-intended officials might leave out. Such prompting resembles the approach suggested by some social scientists who believe that even officials with implicit bias against a group can surmount that bias when prompted to consider issues from another point of view.

Unfortunately, according to the Government Accountability Office, states and local governments have an uneven record in implementing the kinds of protections exemplified in Section 512.[25] For the Stafford Act to live up to its potential for protecting vulnerable populations, Congress should amend the act to include a private remedy under Section 308 as well as a right of action for citizens to force state and local governments to comply with planning requirements like Section 512.

EXECUTIVE ORDER ON ENVIRONMENTAL JUSTICE

In planning for potential disasters, President Clinton's Executive Order on Environmental Justice could provide a template. Recall that this Order requires federal agencies to "address[ ], as appropriate, disproportionately high and adverse human health or environmental effects of its . . . policies . . . on minority populations and low-income populations."[26] Executive orders like these are usually unenforceable and for this reason are ignored unless the White House maintains pressure on the agencies. But a secretary of Homeland Security (or director of FEMA) who is interested in promoting a disaster justice approach could find within this language a charge to develop a more holistic understanding of hazards management. Or a committed president could use this existing order as further reason to insist that agencies working with disaster preparedness provide for the needs of poor communities and communities of color, many of whom, experience shows, will lack transportation, traveling money, child and pet care, Internet access, and the like.

A PROPOSED EXECUTIVE ORDER ON DISASTER JUSTICE

Building on this idea, I would recommend that the White House issue a new executive order—perhaps called the "Executive Order on Disaster Justice"—that would attempt to integrate disaster justice values into the planning of all federal agencies.[27] Patterned after the Executive Order on Environmental Justice, this new order would require federal agencies to consider disaster justice in all policies and activities related to disaster mitigation, response, and recovery. Agencies would be required to identify, address, and protect against conditions that result in disproportionate *or* serious adverse effects on especially vulnerable populations, including minorities, women, children, the elderly, the disabled, the non-English-speaking, the undocumented, and the poor.

Like the Executive Order on Environmental Justice, a disaster justice executive order would not be concerned with an agency's discriminatory intent but rather the *outcome* of its actions or polices, whether or not the outcome was intended. This feature helps avoid injustices that are caused by imperfect economic models, structural bigotry, or implicit bigotry. Unlike the existing Order, the one I propose would require agencies to address adverse effects on vulnerable populations that are *either* disproportionate *or* serious. By requiring that agencies address only disproportionate effects, the existing Order excludes some obviously needy communities while allowing a paralyzing debate about how unequal an exposure or injury has to be before it is worth considering.

At least part of the debate has involved the question of whether it the *exposure* to a risk or pollutant or the ultimate *injury* from it that should be compared among communities. That inquiry is seldom productive. A difference in exposure is not always the sine qua non of environmental injustice. Even if exposure to air toxins is similar in two neighborhoods, one rich and one poor, the degree of inconvenience and harm would probably be higher in the poor neighborhood, where overall health, access to medical care, diet, and other factors will make a difference. But focusing on disparity in injury is not always the right answer either. Measuring differences in injury is usually much more difficult and less certain than measuring differences in exposure. If you have evidence of disparate exposure to a pollutant, it might be too burdensome to require evidence of disparate injury, too. Moreover, a community should be able use evidence of disparate exposure to argue for more effort to prevent disparate injuries from occurring in the future.

The best way around the problem is to allow evidence of disproportionality in terms of *either* exposure *or* injury to count as an indicator of a community deserving of special attention. But the requirement of disproportionality should be dropped where it is shown that a "vulnerable community" (as defined by race, sex, or other relevant characteristic) is bearing (or is likely to bear) "a serious adverse effect." The rationale here is that social vulnerability in the context of environmental dangers almost always makes life worse, whether definitive evidence can be found or not. When protecting a nation's least powerful, it is better to be safe than sorry.

In the typical environmental justice context, which involves chronic pollution, the distinction between "disproportionality" and "serious adverse effect" is troublesome at best. In the context of sudden disaster, the distinction is tragic. Imagine requiring that only vulnerable communities with unequal risk of direct *exposure* be given special thought in the planning process. That policy would do nothing for the wheelchair-bound grandfather on a fixed income whose healthy neighbors (with a car in the garage) are also exposed to the risk of flood. Or imagine a poor Latino neighborhood in the Midwest being denied special consideration in a response plan for severe heat waves because the last time such a disaster struck, white people died at higher rates. In this way, disproportionality (at least as defined by exposure or obvious injury) seems less important in the context of disaster. This may be because in disasters the role of vulnerability is more dramatic and thus more shocking to the conscience. It is also, I think, because disasters are (by definition) relatively rare, making their social consequences harder to predict. Just because Latinos survived Chicago's 1995 heat wave at higher rates than whites doesn't mean they would necessarily do so again. Because we know that certain social vulnerabilities like race and income almost always make it harder to survive naturally triggered disasters, we should err on the side of caution, directing the government's attention to vulnerable populations that risk *either* a disproportionate effect *or* one of a defined level of seriousness.

An executive order on disaster justice should also require FEMA to create an advisory committee charged with developing uniform standards and methodology for agencies to use in carrying out the order's mandate. The lack of uniform standards and procedures made Clinton's order on environmental justice hard to follow and difficult to evaluate. Some agencies, like the EPA, developed regulations intended to guard against discriminatory outcomes; others did nothing. An advisory committee would bring uniformity by defining or at least clarifying special terms like "disproportionality" and "serious adverse effect" to the extent that it is possible to do so (some terms are hard to define, given the diversity of contexts in which they potentially apply). This committee could require all agencies to consider the consequences their actions might have for ease of evacuation, vulnerability of important facilities, and stability of natural barriers like wetlands or forests. Reflecting the value that environmental justice advocates place on participation, the committee would include some representation from vulnerable communities and would seek out their viewpoints through hearings or other outreach efforts. The methodologies adopted would also give communities a substantive role in implementing the order's goals.

The president might also use the executive order to promote a more affirmative agenda for job creation in areas related to disaster preparation and recovery. Again, environmental justice provides the model. For instance, some environmental justice advocates have urged the government to provide incentives to promote the creation of "green-collar" jobs in inner cities and struggling industrial towns.[28] They imagine a new generation of workers who earn good wages and benefits while building light-rail systems, servicing wind turbines, and installing solar panels on neighborhood homes. In furthering the goals of a more environmentally friendly economy, these advocates seek also to benefit those workers whose communities have long been vexed by racism, pollution, and crime.

Following this course, the president could, through executive order, direct agencies to give careful consideration to vulnerable communities when developing workforces to maintain natural and artificial infrastructure and to prepare communities for storms and floods. Most important, agencies could be told to prefer local workers, when feasible, in all cleanup and recovery activities. Certain practices associated with "disaster capitalism," such as the suspension of prevailing-wage laws or unnecessary no-bid contracts to outside firms, could be prohibited or at least discouraged (so long as the directive is made consistent with existing law).

Finally, the new executive order on disaster justice should include procedures for holding agencies accountable for these obligations. The existing Order failed miserably in this regard. Specific goals were never required of agencies, and important terms were vaguely defined. Most important, no single individual or office within the White House was charged with monitoring progress or helping agencies meet the stated goals. The new executive order could take these lessons into account. It could require each agency to develop an agency-wide plan to identify and address actions or programs that put vulnerable populations at special risk, according to detailed and measurable criteria. The plans could also be required to describe an affirmative agenda for including vulnerable communities in the work of restoring infrastructure, preparing for disaster, and recovering from destruction. The plan could be updated on a regular basis. A department within the White House could be charged with vetting the plans and overseeing implementation of the president's goals across all agencies.

One of the strengths of such an executive order on disaster justice would be its general application and flexibility. One can imagine agencies developing plans that facilitate better warnings and evacuations, recognize cultural sensitivities and fears, strengthen the response capacity of nursing homes, and target economic and residential redevelopment in places where people have suffered most. Ideally, all of this would be conceived and implemented with the participation of at-risk communities around the country.

This proposal is not without weaknesses. Some will fear that such a directive would bog government down with even more busywork. Some of the offshoots, say a presumption against no-bid contracts, might also impede an efficient response. Responding to sudden destruction, after all, requires quick decision-making and a clear chain of command. My response is that bureaucracy must engage in the messy work of inclusive and deliberative planning so that when the "Big One" strikes, we can be confident of the marching orders. That work should be done early on, *before* a threat appears on the horizon. When an evacuation is ordered and undocumented residents are wondering whether or not to jump on the city bus, there would be no need for last-minute wrangling between social activists and immigration officials about document checks; U.S. Immigration and Customs Enforcement would already have established the policy in advance. When a storm cripples the nation's refining capacity and the EPA is considering a waiver of air regulations to boost capacity in unaffected facilities located in poor black communities, there would already be a procedure, perhaps with community input, to narrowly tailor such waivers and to monitor their effects. The idea is to use the deliberative process to set a foundation for fast response when it is needed.

An objection from the other side of the political spectrum might be that an executive order, by its nature, does not go far enough. The order's breadth and flexibility ensures that the real substance will be determined by the agencies or (if you trust White House oversight) the president. That works if the administration believes disaster justice is a priority, but not if it doesn't. To make matters worse, there is rarely a way for citizens to enforce the requirements of an executive order. In 2000, an environmental justice challenge to a project at the New Orleans Industrial Canal was rejected when a court found that the original Executive Order on Environmental Justice was unenforceable.[29] In other words, only the White House fox can guard this chicken coop. This objection is real, but it should not be used to dismiss the whole project. Without a directive like this, it would be harder for even the most committed White House or agency to make disaster justice a priority. The mere existence of the order would help raise the issue in the minds of the American public. The Executive Order on Environmental Justice, for all of its weaknesses, did succeed in raising the profile of environmental justice in the eyes of state and local governments, environmental groups, and the press. The order I propose would do the same for disaster management.

COMPENSATION FUNDS

Commentators sometimes raise the possibility of compensatory legislative responses like reparations or special funds as a way of compensating victims for losses that are particularly unconscionable. Farber, for instance, believes there is a "powerful argument" for providing reparations to the families of those killed in the New Orleans flood after Katrina.[30]

Litigation through private rights of action is one way to achieve compensation for vulnerable people who have been put at special risk in times of disaster. Special compensation funds, established by Congress, are another way. Funds like these typically are intended to compensate individuals who have suffered certain described injuries resulting from a natural or technological disaster or, more recently, an act of terrorism. The money is appropriated by Congress and distributed through a no-fault administrative process. As traditionally designed, compensation funds make no explicit distinctions between vulnerable and nonvulnerable populations. But because vulnerable populations are often more likely to be exposed to hazards and because their susceptibility to harm is likely to be higher, and because recovery is generally more difficult, compensation funds are especially relevant to those hoping to build resilience in a disadvantaged community.

Perhaps the best known compensation fund in recent times is the September 11th Victims Compensation Fund, established by Congress 10 days after the attack on the World Trade Center in New York City. Compensation was limited to persons who were present at the crash site and who suffered physical injury or death. The fund did not cover property loss, emotional harm, or risk of future injury. As a condition for compensation, claimants were required to waive their right to file suit against the airlines, the airline manufacturers, the city of New York, or other potential defendants. Kenneth Feinberg, the special master of the fund, issued regulations to govern the claims "that in some instances seemed to go significantly beyond the statutory language."[31] The regulations set a floor of about $250,000 on economic recoveries and provided a cap for recepients with high incomes. Noneconomic losses were valued at $250,000 for each victim and $100,000 for close relatives. Feinberg's intention was to approximate the range of tort compensation so as to encourage eligible claimants to participate. In the end, 97 percent of surviving families made use of the program; the fund paid out $7 billion, with the average payment totaling $1.8 million.

Two lesser-known compensation funds involve Idaho's Teton Dam breach in 1976 and New Mexico's Cerro Grande Fire in 2000. In the first case, a federally constructed dam in eastern Idaho crumbled minutes after it was filled to its intended capacity. Billions of gallons of water swept through Snake River Valley, destroying farmland, livestock, and five downstream towns. Thousands of buildings were lost, and 11 people died. Investigations showed that the earthen dam, which had been designed by the Army Corps of Engineers for the Bureau of Reclamations, had been built on unstable soils and was flawed in both design and construction. The U.S. Department of Justice was prepared to defend against expected lawsuits by invoking government immunity under the Flood Control Act of 1928. Instead, within a week president Gerald Ford requested a $200 million appropriation to start a victim's compensation fund without assigning responsibility for the dam's failure. The legislation, called the Teton Disaster Assistance Act of 1976, allowed all persons who suffered death, personal injury, or loss of property directly caused by the dam failure to receive full compensation, less their recovery from insurance. Through a hastily assembled administrative claims process, 7,500 claims were settled for a total of $322 million.[32]

The Cerro Grande Fire started on May 4, 2000, National Park Service employees set what they planned to be a "controlled burn" to reduce dead growth in the Bandelier National Monument in northern New Mexico. The fire quickly blazed out of control and in a matter of days had become the largest wildfire in the history of New Mexico. The fire destroyed about 43,000 acres, including 7,500 acres belonging to the Los Alamos Laboratory. More than 200 buildings were damaged, and 400 Los Alamos residents lost their homes. In addition, the fire swept through hundreds of waste disposal sites, poisoning acres of land and water with dangerous levels of radioactive material. Two months later, Congress passed a multimillion dollar compensation program intended "to compensate as fully as possible those parties who suffered injuries and damages from the Cerro Grande Fire."[33] The program compensated claimants for personal injury, property loss, reduced real estate values, and even the cost of debris removal.[34]

Against this backdrop, Congress's refusal to seriously consider a compensation fund for the victims of Katrina remains a source of pain in the Crescent City—another sign, for many, that America's social contract stopped at the Louisiana border. The case for a Katrina fund is especially compelling in some ways. Like 9/11, the New Orleans Flood was jarring in both its scope and suddenness. Images from CNN and other broadcasters were seared into the public consciousness, making the destruction of this American city part of everyone's history. Like the Teton Dam breach, the unchallenged evidence is that the flood was caused by defects in design and construction of barriers that were under the supervision and control of a federal agency. (This point distinguishes the New Orleans Flood from the destruction of the World Trade Center, which was caused by a foreign terrorist organization.) Moreover, these engineering mistakes brought destruction on a city with one of the highest concentrations of vulnerable people (minorities, the poor, the disabled) in the country.

On the other hand, certain facts (some substantive, some political) may have worked against a Katrina fund.[35] First, Katrina flood victims were limited in their ability to organize and exert political influence because of their social vulnerability, geographic dispersion, and the fact that state and local government was initially in sheer chaos. The families of 9/11 victims, in contrast, were, by and large, better educated, wealthier, and still living in intact households.

Second, as Farber points out, it is possible that federal lawmakers perceived Katrina victims as being at least partially at fault for their plight. While some who stayed behind had no reasonable means to evacuate, others did. While a many homeowners carried flood insurance (in proportions higher than most coastal cities), thousands did not. Besides, city residents knew the area was prone to flood and, some would say, implicitly assumed the risk.

Third, Katrina victims did not appear to have a reasonable chance of recovering damages from the federal government and its private contractors because of the sweeping immunity provided in the Flood Control Act of 1928. (Still, recovery is theoretically possible: several suits, including *In re Canal Breaches*, mentioned in Chapter 6.) continue to wind their way through the courts.) In contrast, the families of 9/11 victims "had a reasonable prospect of collecting massive tort damages against the airline industry, giving them political leverage."[36] Victims of the Cerro Grande Fire also appear to have had potential legal claims, because the fire was started by a burn "controlled" by park personnel that may have been negligently executed. Fourth, lawmakers may have believed that Congress had already provided enough through other programs, including $8.1 billions of dollars in block grants for Louisiana's home buyout program.

Finally, research in social cognition suggests that lawmakers' attitudes about any of the above elements *may* have been affected by unconscious stereotypes related to ethnicity or skin color—the same kinds of stereotypes that apparently influenced people to short change hypothetical Katrina victims in the surveys conducted by Shanto Iyengar and Richard Morin (Chapter 6).

Disaster relief, or course, should never be constrained because of characteristics like skin color and ethnicity. Nor should the ability to obtain compensation entirely depend on a group's political clout or ability to threaten complicated lawsuits. The point of disaster justice, in fact, is that people who lack influence and leverage are exactly the people who are most in need.[37] Contributory negligence (like *refusing* to evacuate or ensure your home for flood) is a valid concern, but could certainly be worked around by, say, discounting benefits for certain people. The availability of other aid programs is also relevant, although Louisiana's home buyout program did little to help people who did not own homes. Even among homeowners, a recent survey has shown that the state's valuation formulas tended to favor whites over blacks (Chapter 6).

Looking at these examples, from the Teton Dam to Katrina, it is evident we need a less ad hoc way of designing large-scale compensation funds. While it is true that every catastrophe is different, it is also true—as this book argues throughout—that most catastrophes share certain predictable characteristics. Thus Congress might consider legislation that sets the parameters and decision-making process that should govern the development of large-scale compensation funds. At a minimum, the law would define the kinds of catastrophes for which it would apply (in terms of scope, suddenness, or, perhaps, cause) and establish a set of criteria that should be considered in developing a fund (or not developing one). Criteria could include (1) factors related to a community's ability to cope with and rebound from the catastrophe, (2) the availability of other aid packages intended for the same populations, (3) the causal role (if any) played by government actors, and (4) the existence of contributory negligence. The prospect of potential lawsuits must be considered, but the remedy should cut both ways. The existence of widespread liability might justify legislative settlement so that disputes could be quickly resolved and assets (whether public or private) could be justly allocated. But the absence of legal liability could also represent evidence of a population's lack of resilience. Congress could authorize

FEMA or another agency to review eligible requests according to these factors and to issue recommendations. Such a process would still be susceptible to political influence, of course, and Congress could always stray from a recommendation. But the mere existence of such a process would ensure that after every major disaster, the need for compensation would be publicly contemplated and that the same relevant factors would always be openly discussed.

International Agreements

There are many international agreements important to disaster management, including the International Covenant on Civil and Political Rights, the Convention on the Elimination of all Forms of Racial Discrimination, and the Protocol Relating to the Status of Refugees, all of which the United States has ratified.[38] Relevant treaties that the United States has not ratified include the International Covenant on Economic, Social and Cultural Rights, the Convention on the Elimination of All Forms of Discrimination against Women, and the Convention on the Rights of the Child.[39] These agreements are more established but were not written with disaster management specifically in mind. While aggrieved individuals have the right to lodge complaints with the United Nations, none of these conventions creates individual rights that are legally enforceable in the traditional sense.

Another agreement relevant to disaster justice is the U.N. Framework for Climate Change, of which the Kyoto Treaty is one aspect. For now, I will focus on two other more recent international agreements that represent attempts to address international disasters in innovative ways: the Hyogo Framework for Action and the U.N. Principles on Internally Displaced Persons.

THE HYOGO FRAMEWORK FOR ACTION

In January 2005, government officials from around the globe met in Kobe, Japan, to discuss disaster preparation and response. By odd coincidence, only days before, an enormous earthquake off the west coast of Sumatra had triggered the devastating Asian Tsunami, which raked the coasts of 11 countries and killed more than 225,000 people. At that international meeting, called the World Conference on Disaster Reduction, all member states of the United Nations agreed to reduce disaster loss by strengthening the resilience of nations and communities at risk. One product of the conference, reported widely by the press, was a plan to establish an international early warning system to identify coastal storms and tsunamis in time for people to evacuate. A related, and even more ambitious product, was an international framework to guide all the world's disaster initiatives through 2015.

The Hyogo Framework for Action, as it is called (named after Japan's Hyogo district), aims to reduce the "human, social, economic, and environmental losses" of natural and technological hazards by strengthening disaster response programs and integrating disaster planning into crosscutting areas like public health, education, and economic development.[40] The Hyogo Framework is a remarkable platform for advocates of disaster justice. The Framework represents the belief that good disaster management (like good environmental management) depends on strengthening the capacity of poor and disadvantaged communities. Its first strategic goal, for instance, links the strategic goal of reducing world risk to a more ambitious effort to reduce world poverty. Recognizing that disasters strip vulnerable groups of hard-won economic gain, the document also urges that disaster management strategies be "mainstreamed" into future multilateral assistance programs. The United Nations notes that "particular attention" should be paid to the African continent, Least Developed Countries, and "[s]mall island developing States."[41] Like the environmental justice movement, the Framework emphasizes within its strategic goals the importance of community organizing and the inclusion of women in leadership and decision-making.

In addition, the Hyogo Framework explicitly recognizes the value of natural infrastructure by incorporating environmental protection, particularly as it applies to food security, resource conservation, and land-use planning. Indeed, the document notes that proposed outcome measurements should conform with international principles of sustainable development, on the grounds that disaster reduction and environmental quality are "mutually reinforcing objectives."[42] Climate change presents an obvious challenge in disaster reduction, and the framework encourages initiatives designed to adapt to a future threatened by rising tides and prolific storms. The Hyogo Framework thus recognizes both the "Go Green" and "Be Fair" agendas advanced this book. (The Framework also deals with risk management, but as we'll see in Part III, it departs from my "Keep Safe" precautionary agenda.)

As you might suspect, nothing in the Framework is binding. Rather, these quasi-legal guidelines are what international lawyers call "soft law": an expression of agreed-on international norms that is intended to influence regional and national policy. To implement these norms, the United Nations has established a multistakeholder task force, the International Strategy for Disaster Reduction (ISDR) System, which helps coordinate public and private initiatives for understanding and reducing disaster risk. The ISDR System also promulgates standards and examples of "good practices" to encourage effective disaster management throughout the world.

As of 2007, more than 38 countries have established "national platforms" for implementing the Hyogo Framework's strategic goals. More than 100 countries now have some agency or other body responsible for implementing and monitoring progress toward these goals. Hyogo-inspired risk reduction strategies have already been designed or implemented in the Andes region, Central America, the Caribbean, the Pacific, Asia, Africa, and Europe.[43] In earthquake-prone Kazakhstan, for instance, the U.N. Development Programme is helping to incorporate disaster response information in grade school curricula as a way of building resilience in poorer communities. Nigeria, whose people sometimes endure heavy rains and dangerous floods, has integrated risk reduction into its regional planning budgets and antipoverty initiatives. In Vietnam, a special partnership of government offices, private donors, and nongovernmental organizations is promoting disaster management for the country's storm-swept jungles and shores; the effort focuses on *both* lifting villagers out of poverty *and* protecting river basins, swamps, and coastal beaches. At the international level, the World Bank, in collaboration with the ISDR System, recently launched a major aid initiative to help "low- and middle-income countries" integrate risk reduction strategies in their development programs.

What do these projects look like on the ground? As just one example, consider Bangladesh. Bangladesh is now engaged in an enormous project to line a third of its shoreline with mangroves to protect it from raging cyclones. When tropical storms cause floods, *two-thirds* of the country can go underwater. Improvements in warning and evacuation systems in the 1990s have saved literally millions of lives. But evacuation cannot save the homes or the cropland; and when those assets are lost, millions of families fall further into poverty. In Gaibandha, a remote part of Bangladesh located at the confluence of the Tista and Brahmaputra rivers, dangerous floods come not from the coast but from these two mighty rivers, whose banks overflow in the wet monsoon season. The residents of Gaibandha have a very hard life. Most live without access to safe water, sanitation, or other basic services.

Often people live along the riverbanks to take advantage of more fertile farmland; but when the monsoons come, the river banks erode, crops wash away, and thousands are left homeless. With no means of earning income, the men search for work elsewhere. The women and children stay behind, where they are plagued by bandits, indentured labor, malnutrition, and early pregnancy.

A nongovernmental organization called Practical Action Bangladesh, along with five local partners, is now piloting a program to encourage settlement away from floodplains and to train adults to earn income in other ways. The program helps communities develop multipurpose shelters, "cluster villages" for resettlement, and housing that is resistant to flood. Residents are introduced to "flood-friendly" cultivation techniques and trained to run small businesses. The ISDR System reports that the project, which targeted 20,000 households, has raised the area's average income and "reduced significantly" the migration of men.[44] The ISDR System hopes to replicate this and many other successful projects in the near future.

Despite small success stories like this, the overall global challenge is enormous. Disasters claim the lives of hundreds of millions every year and cost the world tens of billions of dollars. Climate change, coupled with sprawling human development, ensures that these numbers will go up. The world's governments have not even begun to contemplate deploying the resources we might ultimately need to stem the tide. But when governments become more serious, we will find that legal frameworks like Hyogo have set the foundations for organizing and developing intelligent ways to reduce the world's exposure and vulnerability to environmental disaster.

As a start, rich countries should make good on their intention to provide at least 0.7 percent of their gross national income in international aid, making sure that disaster risk reduction based on the Hyogo Framework is a key part. Development assistance should also mitigate underlying social risks by improving education, health, and water and sanitation systems and by fighting discrimination against women, minorities, and lower-caste people.[45]

THE GUIDING PRINCIPLES ON INTERNALLY DISPLACED PERSONS

Days after Katrina, when it became clear that the 2 million residents who had fled their homes would not be returning for awhile, Americans wondered what to call them. Some news organizations referred to them as "refugees." But many Louisianans resented that term. They pointed out, correctly, that in legal parlance the word *refugee* refers to a person who has fled to another country, not just crossed a state line. Given the government's insensitive response to their rescue, calling them "refugees" reinforced the view that individuals—many of whom were urban, poor, and black—were not part of the "real" America. So the press and the public settled instead on "evacuees," "storm victims," or, in casual conversation, "Katrina people."

What many did not know was that the United Nations, whose capacity for inventing new terms is unlimited, had already established a phrase to describe people who have been forced to flee their homes, for reasons of armed conflict or disaster, but who remain in their country: "internally displaced persons" (IDPs). While numbers are wildly uncertain, experts believe there are roughly 24 million IDPs in the world at any given time. Most are probably displaced for reasons of war or other violent conflict. But disasters, both natural and humanmade, contribute significantly to the problem, as in Ethiopia (drought), Pakistan (earthquake), India (flood), and China (earthquake). As we saw in the example of Myanmar after Cyclone Nargis (Chapter 5), the internally displaced are at special risk of being neglected or mistreated for reasons of ethnicity, sex, age, or other characteristics.

Unlike international refugees, international law offers no special protections for IDPs. Moreover, the United Nations scatters responsibility for IDP issues across many agencies, making comprehensive solutions more difficult. The United Nations has endorsed, however, a document entitled Guiding Principles on Internal Displacement, assembled at the request of U.N. secretary-general Boutros Boutros-Ghali in 1992.[46] Like the Hyogo Framework, the Guiding Principles are another attempt to reduce risk by attacking oppression and set a sensible foundation for future legal development. While not binding, they are said to reflect "international human rights and humanitarian law and analogous refugee law."[47] Some provisions concerning physical security and the right to travel, for instance, are drawn from the Covenant on Civil and Political Rights. The 1992 document details, in 30 principles, the guarantees that nations owe to IDPs in all three phases of internal displacement: predisplacement, actual displacement, and return or resettlement.

Several of these provisions could be characterized as "process-based" antidiscrimination rules. In this way, they represent a set of negative liberties meant to expand the bounds of membership in the "social contract," as

Ignatieff might describe it (Chapter 7). For instance, the Guiding Principles state that IDPs enjoy "full equality" with other persons in the country.[48] They hold regardless of one's legal status and must be applied "without discrimination of any kind, such as race, colour, sex, language, religion or belief, political or other opinion, national, ethnic or social origin, legal or social status, age, disability, property, birth, or on any other similar criteria." This may require that special "protection and assistance" be afforded to certain groups such as women, expectant mothers, children, the elderly, and the disabled. For instance, governments are instructed to give "special attention" to the health needs of women, including "reproductive health care," and to make "special efforts" to include women in the planning and distribution of basic supplies such as food, water, clothing, and shelter. In addition, governments owe a "particular obligation" toward indigenous peoples and others who may have a "special dependency on and attachment to their lands."

The larger portion of guarantees sets out a list of positive and negative liberties meant to define the social contract being proposed. In this way, the Guiding Principles resemble Joseph Singer's social relations approach to property rights (Chapter 4), but here in the context of public law. That is, the Guiding Principles seek to define the minimum level of resources an IDP would need in order to pursue a safe, purposeful, and dignified existence (given the extreme circumstances of war or disaster). Then the principles create entitlements to those resources and integrate them into the relationship that all inhabitants (for we have moved beyond legal residency or citizenship) have with a host government.

To this end, the General Principles hold that nations have the "primary duty and responsibility" to protect and provide assistance to IDPs. When aid or assistance is offered from foreign governments or outside nongovernmental organizations, it should be considered in good faith and not refused unreasonably. In addition, IDPs are protected from "arbitrary" efforts to relocate them that are not sufficiently related to safety. Thus governments may not use war or disaster as a mere excuse to segregate populations, punish minority groups, or free up valuable land for large-scale development. Displaced persons are entitled to shelters that meet "satisfactory conditions of safety, nutrition, health and hygiene" and do not require the separation of families. Those in shelters always have the freedom to leave. An individual also enjoys the "liberty of movement," the freedom "to seek safety in another part of the country," and the freedom "to choose his or her residence." Governments also owe IDPs the "conditions and means" to return to their place of former residence or to resettle in another part of the country. Authorities must also help IDPs recover their property and possessions. If such recovery is impossible, "competent authorities shall provide or assist these persons in obtaining appropriate compensation or another form of just reparation."

These goals are impressive and ambitious. The Guiding Principles carry forward the theme expressed in the Hyogo Framework and in the American environmental justice movement that protecting people from future harm means reducing their social vulnerability and strengthening their social resilience. The Guiding Principles also skirt the issue of discriminatory intent—so vexing in the environmental justice context—by establishing substantive rights for special needs that some groups are likely to have. Thus we do not have to debate whether the failure to provide prenatal care at a long-term shelter represents an unintentional slight or intentional discrimination against women: either way it is a violation of Principle 19 (the guarantee of reproductive health care).

The Guiding Principles draw obvious lessons from (and in some cases anticipate) examples of disaster injustice that I have already reviewed. The antidiscrimination provisions, for instance, would apply to the ethnic bias some perceived in the Myanmar evacuation (Chapter 5). The protections against arbitrary removal for purposes of economic development recall Klein's investigation of the redevelopment of Arugam Bay after the Asian Tsunami (Chapter 6). It's hard to read the guarantee of "satisfactory conditions of safety, nutrition, health and hygiene" contained in Principle 7, without thinking of New Orleans's hellish, ill-equipped Superdome. Similarly, "the right to seek safety in another part of the country" evokes the eerie standoff between stranded evacuees and police officers on the Gretna Bridge (Chapter 6).

Still, many of the provisions are vague and raise difficult questions that may hurt their chance of one day being incorporated into binding law. Take the blanket prohibition of discrimination. Should Pakistan, say, be able to segregate relief camps by sex in order to comport with religious custom or protect women from violence? If so, should India be able to segregate relief camps by caste for similar reasons? The answers are not clear and probably depend heavily on context. Would the protection of those without "legal status" bar American immigration officials from checking residency documents when evacuees return to their cities, as reportedly happened after Hurricane Gustav in 2008? And what of the requirement that governments provide people with compensation or reparations for lost property or assist them in recovering it? Read broadly, this right could reverse the West's traditional concept of sovereign immunity (which holds governments immune from civil law suits unless consented to) and the modern idea of regulatory discretion (which allows government latitude in protecting the public safety). Surely, governments would want some explanation or limitation on this provision before accepting it as binding law.

These uncertainties notwithstanding, the Guiding Principles are gaining traction and are presumably making a difference in people's lives. Several U.N. entities, such as the U.N. Development Programme, the High Commissioner for Human Rights, and UNICEF, have incorporated the Guiding Principles into their actions and policies. Some of the United Nations treaty bodies that monitor the implementation of U.N. human rights conventions have referred to the Guiding Principles in their observations of member states. While the U.S. government does not recognize the Guiding Principles as law, the U.S. Agency for International Development explicitly uses them to guide foreign assistance efforts in regions of disaster or war. A few other countries, including Burundi, Colombia, the Philippines, Sri Lanka, and Uganda, have integrated some of the principles into their own national policies.

The next step in international law requires that a set of *binding* principles be negotiated to provide some minimum protection for vulnerable populations in times of catastrophe. Bahame Tom Nyanduga, a special rapporteur for the African Commission on Human and Peoples' Rights, has called the absence of binding protection for IDPs "a grave lacuna in international law."[49] These binding principles need not be as elaborate as those in the nonbinding Guiding Principles document, but they should be clearly defined and available to all. The industrialized countries need to address another gap by committing to a binding scheme for providing funds and technology to developing countries to promote sustainability initiatives of the kind envisioned in the Hyogo Framework. The most logical move would be to incorporate such a scheme within the United Nations Framework on Climate Change, since many of the natural hazards of concern in Hyogo will be amplified by global warming and sea-level rise.

The Lakes of Pontchartrain

In Part II, I have tried to lay out the principles for the directive I call "Be Fair." To be fair requires that government and the public first acknowledge that naturally triggered disasters are *not* social equalizers. Vulnerable groups often suffer much more than others in the aftermath of floods, earthquakes, and other catastrophes, whether in industrialized democracies or agrarian police states. They always have, and unless we change course, they always will. For many of the same dynamics that drive inequalities in peaceful times—market systems, corruption, and bigotry—go into overdrive when crisis hits, and the traditional checks on these social forces break away and fall apart.

In this part, I have borrowed from the still developing environmental justice movement to suggest both a critique of the current system and a set of hopeful reforms. The critique accepts the notion that everyone is entitled to some basic package of protection from natural hazard and that such protection depends on remedying the social vulnerability of populations encumbered by poverty, disability, discrimination, and other personal characteristics. Remedying the problem requires shoring up traditional protections for vulnerable people and implementing structural reform so as to discourage bigotry injustices based on bigotry. It also requires special laws to address the particular issues that arise in disasters like evacuation and internal displacement. Part III, which features an analysis of precaution and risk management, will occasionally remind us to recall the lessons of fairness in order to make sure that risk is not being borne by some unjustly.

The day after Katrina's surge rushed into the Pontchartrain estuaries and burst the levees, a friend sent me a consoling e-mail introducing me to "The Lakes of Pontchartrain," an English-language Creole ballad thought to date back to the early nineteenth century.[50] The song tells the story of an impoverished foreigner, perhaps a British soldier, alone at night in what were then the swamps abutting Lake Pontchartrain. Exhausted and desperate, he comes upon a beautiful Creole woman.

The man says:

> My pretty Creole girl, my money here's no good,
> But if it weren't for the alligators,
> I'd sleep out in the wood.

Then she says:

> You're welcome here kind stranger,
> Our house is very plain.
> But we never turn a stranger out,
> From the lakes of Pontchartrain.[51]

The verse suggests an important moral: in times of crises, good people take care of one another, and this is true whether your house is fancy or plain. To "be fair" means to stay informed about the needs of others and to treat them as your own. Even in howling winds, it means you never bar the door.

III

KEEP SAFE

8

PRECAUTION AND SOCIAL WELFARE

The map is not the territory.

—ALFRED KORZYBSKI

After Katrina I studied hundreds of maps of my city and its surrounding coastal features: thermal maps of twisting winds; maps of erosion, elevation, and housing patterns; maps pinpointing survivors climbing through roof-holes and into helicopters; maps of the dead. But none of them told the true story of how more than 30 billion gallons of water swallowed a city. I think of the ambitious cartographers imagined by Jorge Luis Borges in his story "On Exactitude in Science."[1] Obsessed with precision, they draft a map of their empire as big as the empire itself—one that coincides "point for point" with the surveyed terrain. But on such a scale, their handiwork is pretty much useless. Eventually the map is abandoned to the erosive forces of sun and ice, its tattered remains inhabited by "Animals and Beggars." The moral is clear: more than maps that show us everything, we need maps that show us what we need to know.

As suggested by my opening quotation from the linguist Alfred Korzybski,[2] maps can be both conventional and metaphoric. We use maps as a way to compress and communicate wide landscapes of information. Some maps, like the widely distributed satellite images of the New Orleans flood, are of the conventional sort.[3] They seek to convey information "point by point," albeit on a smaller scale, and thus suggest a form of objectivity or "literalness." Still, we are aware that their perspectives, their digital coloring, even their abstract visual appeal affect what we see—and *don't* see. You will find no blood on a satellite image of storm-wracked New Orleans, for instance. And you have to imagine the worried brows. Other "maps" were supposed to show us those things. Generations back, engineers and policy-makers developed sophisticated models and formulas designed to capture the probable forces of future storm winds, water surges, and environmental destruction in New Orleans in the event of a storm like Katrina. Other mapmakers—the economists and risk assessors—plotted, with terrier-like resolve, elaborate templates designed to predict, in some rough way, the danger to residents, the preferences of citizens and consumers, and the demands of the public good. The environmental predictions were based on woefully inadequate information; the risk assessments were cramped and narrow. Nonetheless, these maps offered a reassuring *illusion* of certitude, set in a filigree of ornamental numbers and graphs.

The conventional maps (like satellite images) show what happened when the Crescent City's levees gave way to an impressive, though predictable, surge. The other maps (the environmental predictions and risk assessments) show why. In a nutshell, we didn't adequately consider how much we didn't know about a Gulf hurricane threat, how spare our climate data were, and how thin was our grasp of public risk. This was not because we didn't *know* the extent of our ignorance. We didn't *see* it. The inherent uncertainty built into our models was not drawn prominently enough onto our maps. Where cartographers of old once emphasized the limits of their knowledge by filling blank corners with flying beasts and colorful serpents, today's conceptual mapmakers sketched only calm seas.

Edward Tufte, a pioneer in "information design," was one of the first scholars to comprehensively study the communication of information and abstraction in map-like images.[4] It was he who famously observed that while the multicolored commuter map for London's Underground was ridiculously inaccurate in geographic terms, it was brilliant in realizing the goal of getting people on the right train with seconds to spare.[5] Tufte would later link the 2003 Space Shuttle *Columbia* disaster with the imprudent use of PowerPoint slides to inform NASA's top decision-makers of the risks created when a piece of foam insulation broke off during liftoff and hit the shuttle's left wing.[6] PowerPoint's coercive bullet-point hierarchy and limited slide area had the effect, he argued, of sifting out uncertainties and downplaying the danger *Columbia*'s crew would face on re-entry. In handling disaster, the medium does, indeed, shape the message.

I use the injunction "Keep Safe" to describe the exercise of precaution. That means adding a margin of safety to the decisions government makes. It means describing the risks and uncertainties of disaster in ways that help us see what is important so that we can make responsible, transparent decisions. This chapter looks at two competing methods policy-makers use to assess risk and uncertainty: the precautionary approach and the social welfare approach. I have argued elsewhere that the social welfare approach, which includes cost-benefit analysis, is an inherently flawed tool when used to craft or evaluate laws protecting health, safety, or the environment.[7] In the case of disaster policy, these flaws are magnified many times. Indeed, as I will show, many otherwise reliable proponents of the cost-benefit model back away from that precipice in the face of catastrophic risk. The precautionary approach, used in most laws governing health, safety, and the environment, is a better method but must be further developed for our purposes.

In Chapter 9, two aspects of the New Orleans flood will give life to this abstract debate: the federal government's choice of storm protection for the region and the influence that a precautionary environmental analysis had on the Army Corps's choice of the city's flood control system. In Chapter 10, I combine the lessons of Chapters 8 and 9 to propose a better way of analyzing risk and uncertainty in cases of disaster where the probabilities are low but the potential harm is great. This model is based on a method that the Corps is, in fact, now using in its analysis of flood control in the Gulf, a model based on scenario planning.

Mapping the Worst Case

Since the 9/11 attacks and Hurricane Katrina, scholars in many disciplines have urged government and citizens to start taking catastrophe more seriously. Referring to nightmare scenarios like global pandemics and asteroid collisions, federal appellate judge Richard Posner warns that "[t]he number of extreme catastrophes that have a more than negligible probability of occurring in this century is alarmingly great, and their variety startling."[8] Even the regional, run-of-the-mill disasters should send a shiver down your spine. Lee Clarke, a social scientist who has devoted a career to studying disaster and terrorism says we don't think about worst cases enough: "Worst case thinking hasn't been given its due, either in academic writings or in social policy. We're not paying enough attention to the ways we organize society that make us vulnerable to worst cases. We're not demanding enough responsibility and transparency from leaders and policy makers. I am not an alarmist, but I am alarmed."[9]

Faced with an alarmingly wide range of worst-case scenarios, what's an alarmed nonalarmist to do? Two models for analyzing hazard have gained wide traction in American law and policy-making. One, the precautionary approach, grew out of the environmental and worker safety movements of the 1970s. The other, the social welfare approach, evolved from public works planning in the 1930s and was revitalized decades later. Adherents of both models have something to teach about worst-case scenarios, though the discourse between them reminds one of an elegantly choreographed knife fight.

Precaution

The precautionary principle holds that where a process or activity threatens harm to human health or the environment, preventative measures should be taken even if some cause-and-effect relationships are not fully understood. There are many versions of the idea, but all emphasize the need to respond to low-probability, high-stakes scenarios against a background of scientific uncertainty. The principle is also attentive to the interests of future generations, a concern especially relevant to disasters that statistically occur only once or twice over a broad time span. Thus a prominent environmental protection act in Australia holds: "If there are threats of serious or irreversible environmental damage, lack of full scientific certainty should not be used as a reasoning for postponing measures to prevent environmental degradation."[10] The U.N. Framework on Climate Change (UNFCC) fashions the principle this way: "Where there are threats of serious or irreversible damage, lack of full scientific certainty should not be used as a reason for postponing [regulatory] measures, taking into account that policies and measures to deal with climate change should be cost-effective so as to ensure global benefits at the lowest possible cost."[11]

In American law, you will find the precautionary principle embodied in at least two kinds of regulatory standards—feasibility standards and those based on open-ended balancing. In contrast, the cost-benefit approach, to be discussed shortly, is an outlier as legislative standards go. "Only two of twenty-two major health, safety, and environmental statutes rely on a cost-benefit test as the statutory standard."[12] The feasibility principle generally demands the maximum level of protection that can be achieved by available technology unless the cost of that protection would threaten the financial integrity of a regulated industry. It is a directive to "do the best you can."[13] While it emphasizes safety, feasibility explicitly considers cost, thus establishing an outer limit to the precautionary drive. In this way it resembles the precautionary statement presented in the UNFCC, which is also concerned with cost-effectiveness.[14] In the United States, feasibility standards govern workplace safety, hazardous air pollutants, point-source discharges into waterways, and many other aspects of modern life.[15]

The second most common form of precaution in American statutes employs a method called "open-ended balancing." In this approach, agencies consider a variety of qualitative and quantitative factors without converting them into to any universal currency (like dollars) or without even assigning them relative weights. The Federal Insecticide, Fungicide, and Rodenticide Act, which governs the licensing of pesticides, provides an example of such a method. Under that law, the EPA is directed to regulate licensing so as to avoid "unreasonable adverse effects on the environment," which are defined as "any unreasonable risk to man or the environment, taking into account the economic, social, and environmental costs and benefits."[16] The financial cost of regulation is considered but is not directly pitted against monetized benefits. Open-ended balancing is precautionary in that it "tilts a statute in favor of increasing protection . . . because an agency is authorized to regulate even if it does not have sufficient information to quantify all of the benefits and prove that they exceed the costs."[17]

Critics of the precautionary approach charge that it is vague, paralyzing, and alarmist. It is vague because one cannot always tell on the face of a situation which way a decision should come out. Vagueness invites politicking and can allow an unelected bureaucrat to substitute her value judgments for the public interest. The vagueness charge may be particularly true for statements like the one in the Australian act quoted earlier (though other parts of that law elaborate on the standard) or in open-ended balancing tests that provide no weights or priorities. The charge seems weaker when leveled against the feasibility approach, which has been construed by courts in a rather uniform way.

Precaution is said to be paralyzing because a precautionary approach, stripped down and taken to its logical conclusion, would forbid all actions and even inaction—every option, after all, carries some microscopic risk. Without more direction, there is no way to know which option and which risk to prefer. A ban on one known toxin, for instance, might lead to a replacement product that is even more deadly. In addition, money spent preventing one type of harm might be better used to create jobs or increase national wealth. This "killer substitute" argument seems overblown, though. Statistically, it makes better sense to control the thing you have some reason to believe is dangerous than to fear future alternatives that you have no information about at all. Precaution advocates are also skeptical that increasing wealth is superior to saving lives, avoiding injuries, or protecting ecosystems and are distrustful of the economic tools used to compare the two choices. They also point out that there is little evidence to support the view that environmental regulations inhibit economic growth or the accumulation of wealth.[18]

Finally, when environmental or safety regulations are driven by precaution informed by public opinion, critics believe, regulations will go farther than they should. This is because individuals have cognitive biases that sometimes focus their fears on some risks and away from other perhaps more deadly ones. People worry much more about, say, contaminated landfills and bioterrorism than probability analysis suggests they should. That's because we are more likely to think about headline possibilities that are easily recalled or, as psychologists say, "available." People also exhibit biases against loss (as opposed to avoided gain) and experiences they perceive as especially dreadful. We've seen cognitive biases before. In Part I, a cognitive bias prevented people from acknowledging the connection between "free" ecosystem services and the market economy. In Part II, cognitive bias helped explain unconscious preferences linked to skin color, race, and ethnicity. Advocates of precaution respond by arguing that even if public outcry brings about profligate spending to prevent, say, leaking landfills, there is no reason to think that that money would otherwise be spent on something more optimal, say, workplace safety. Besides, such biases may reflect cultural values that, while not strictly utilitarian, are deserving of government's recognition and respect.[19]

Social Welfare and Cost-Benefit Analysis

Social welfare theory (welfarism) strives to maximize the overall welfare of society, where welfare is defined as the aggregate utility of all members of society. Utility in this sense refers to individual satisfaction or pleasure. Strictly speaking, the theory does not recognize the relevance of overarching "social" values beyond those held by individuals. Thus values like fairness or equitable distribution are relevant only insofar as individual members can be said to value them. Having evolved from the works of Jeremy Bentham and John Stuart Mill, welfarism is neoclassical in its economic outlook. It has been embraced more recently by neoliberals, as discussed in Parts I and II. Neoliberalism's emphasis on the individual economic actor, you will recall, slowed American property regimes' acknowledgment of group interests in natural infrastructure (Chapters 1 and 3). Neoliberalism also helps justify "risk avoidance" in environmental protection and "you're on your own" emergency response strategies (Chapters 5 and 6).

Welfare advocates usually recommend cost-benefit analysis as the primary way to enhance social welfare. "Cost-benefit analysis" refers to a regulatory process in which regulators try to "add up the benefits of a public policy and compare them to the costs."[20] To compare these nonmonetary benefits to the monetized costs of a government project or regulatory compliance, analysts assign dollar amounts to these benefits. These dollar values are often based on wage premium studies that purport to measure in some broad way a person's "willingness to pay" for a benefit or to be relieved of a burden. To monetize the reduced risk of a fatality, for instance, an economist might examine the "wage premium" a blue-collar worker earns for working at a risky job.[21] If a worker generally earns an extra $600 per year in exchange for a 1-in-10,000 chance of being killed in an accident on the job, the economist concludes that the aggregate value of one "statistical life" is $6 million—which, as it turns out, is just shy of the $6.1 million figure the EPA used in reviewing its standards for arsenic in drinking water.[22] In this sense, a "statistical life" refers not to an actual life but to a mental place card the analyst uses to signify the presence of collective risk.[23]

Because health and environmental regulations often seek to prevent harms in the distant future, cost-benefit analysis for such regulations include a discount factor to reduce into present dollars the value of future benefits, whether or not they are monetary in nature. (Costs incurred over the long run are discounted, too.) The goal is to create a ledger of "commodified" interests that can be traded for one another so as to enable the analyst to optimize the social benefit of regulation.

The economic roots of the cost-benefit approach are anchored in social welfare theory and liberalism. The political roots of this approach are anchored in positivism: the theory that knowledge is objective and that social inquiry should be limited to facts that can be directly observed.[24] Cost-benefit analysis, like all positivist methods, aspires to separate facts from values, holding that "knowledge accumulation should occur independently of the researcher's preferences or expectations."[25]

The cost-benefit approach is used in some aspects of natural-hazard reduction. Federal agencies use cost-benefit analysis in designing levees and dams, for instance. Some have urged its use in shaping policy on global warming, perhaps the biggest disaster-generating machine we have ever seen. The origins of cost-benefit analysis go back to a French engineer named Jules Dupuit, who in 1848 proposed a utility-based means for evaluating public works projects. The first use of the approach in the United States is generally attributed to the Army Corps of Engineers, when it found itself, after the Great Mississippi Flood of 1927 and the Northeast Flood of 1936 charged with the responsibility to protect against catastrophic flooding along all navigable waters of the United States (Chapter 4). In an effort to prevent waste and discourage pork barrel politics, the authorizing legislation, the Flood Control Act of 1936, permitted projects only where "the benefits to whomsoever they accrue are in excess of the estimated costs."[26] This requirement had already been Army Corps policy, but now it was cemented into American law. This early version of cost-benefit analysis was different from what most policy-makers use today. Early analyses, for instance, did not always share the same principles or techniques. And, significantly, they did not seek to monetize intangibles like saved landscapes or saved lives. To this day, the Army Corps does not factor human safety into its cost-benefit analyses.[27]

Modern cost-benefit analysis came to Washington by way of Detroit, with Robert McNamara and the so-called Whiz Kids, a group of 10 World War II veterans who started out as young executives at Ford Motor Company and eventually introduced the management science they learned in the air force to all of American business.[28] What came to be known as management science emphasizes the use of math and statistics in making decisions. At Ford in the 1950s, McNamara helped develop new methods of financial analysis based on cost-benefit models. When he became secretary of defense in 1961, he brought his ideas to Washington, where he applied cost-benefit reviews to some Department of Defense activities. Presidents Nixon and Carter modestly extended cost-benefit reviews to other regulatory actions, but things did not really get cooking until the "Reagan Revolution." In 1981 Reagan issued an executive order that (1) required "regulatory impact analysis" of all "major" rules (usually rules that would impose more than $100 million in compliance costs), and (2) resisted agency adoption of such rules unless benefits "outweigh[ed]" costs.[29] Under this and other orders, a system of comprehensive regulatory review was established in which the White House Office of Management and Budget (OMB) would supervise the implementation of most federal laws, including those governing worker safety, food and drugs, consumer products, and the environment. If the OMB believed that an agency had drafted a regulation whose compliance costs exceeded the safety benefit, it could use its influence to persuade the agency to revise its proposal.

President Reagan used this regulatory review process to impede agency rule-making. President George H. W. Bush also used the process as an antiregulatory device.[30] President Clinton took the centralized powers of the OMB and made them his own. He replaced Reagan's executive order with a new one advising that benefits should instead "justify" costs, softening the antiregulatory effect.[31] When President George W. Bush inherited the Clinton executive order, he retained the language but employed it more aggressively to resist new regulations. President Obama, like Clinton, seems interested in fine-tuning the cost-benefit approach, but there is no indication that he plans to significantly change how cost-benefit analysis is used in centralized regulatory review.

The migration of the cost-benefit approach from public works projects to the regulation of health, safety, and the environment was enabled by innovations in economic analysis involving theories of optimality and, more important here, the use of willingness-to-pay as a means of valuing improvements in morbidity, mortality, and the preservation of natural resources.[32] Today advocates of cost-benefit regulatory review characterize it as a method whose necessity was once debated but is now established; the only question left is how best to implement it.[33] A growing chorus of precaution advocates, many coming from the legal academy, insist that a victory lap is premature.[34] Objections to this kind of regulatory review involve both substantive and procedural arguments. The substantive argument is that cost-benefit analysis is an improper method for evaluating important regulatory protections that involve health, safety, and the environment because it is inherently incapable of measuring putative nonmonetary benefits in any useful way. In addition, the substantive argument goes, even if such measuring were theoretically possible, the malleability and complexity of the calculations make it too easy to smuggle in an antiregulatory bias. (Of course, one could also plant a proregulatory bias; but in practice, cost-benefit review has almost always cut *against* regulatory action.)[35]

The procedural argument, which is independent of the substantive argument, is that such emphasis on cost-benefit review is inconsistent with congressional intent. Hardly any statutes protecting health, safety, or the environment require or even invite cost-benefit evaluation in shaping regulation. Indeed, some statutory provisions, such as those governing ambient air quality, forbid it.[36] At worst, an executive order that advises against environmental regulation unless benefits outweigh costs undercuts a congressional directive. At best, cost-benefit review is a sideshow, a review of statistical gymnastics that Congress had always refused top billing. A second variation on the procedural argument, also involving the executive and legislative branches, says that such a comprehensive regulatory review process (whether or not based on cost-benefit analysis) unduly centralizes executive powers. It is one thing to have the president's cabinet appointees (or other chosen officials) directing policies within the government's many scattered departments; it is another for all those policies to be so tightly managed by a set of OMB officials whom most Americans have never heard of. Such concentration threatens the discretionary authority of agency heads and thwarts the intent of Congress, whose laws properly delegate such authority to the agencies.[37]

Applying this history, we can see that cost-benefit analysis affects disaster policy in two important ways. First, agencies like the Army Corps use cost-benefit tools to design protective intrastructure like dams and levees. If the designers use models that exclude important benefits, such as lives saved or ecosystems preserved, their proposals will be inadequate. Even if their models include such nonmonetary benefits, their use of monetization and discounting will cause many benefits of protection to be underestimated.

Second, and equally important, because cost-benefit analysis is incorporated into centralized regulatory review, its methodology currently affects how *every* significant environmental, health, and safety rule is evaluated. Consider natural infrastructure. Nearly *all* of the regulations that protect resources important to disaster risk reduction—from coral reefs to cypress swamps to old-growth forests—encounter at some point the cost-benefit sentinels at OMB. For example, recall how the second Bush administration justified its neglect of the Clinton administration's "roadless rule" by arguing that the benefits of preserving ancient forests were not worth the costs (Chapter 4). Cost-benefit review similarly has an extensive influence on pollution and safety regulations important to disaster justice. For instance, a proposal to apply federal waste storage standards to coal-ash slurry—which flooded the Tennessee neighborhood discussed in Chapter 7—would undergo cost-benefit review. Such review would also likely be required for stricter regulations governing the storage of petroleum products in floodplains, such as those released during Hurricanes Katrina and Gustav.

What do precaution advocates say about welfarism and cost-benefit analysis? Precaution advocates are not against all welfare principles. When precautionary initiatives require that measures be "cost-effective" (as in the UNFCC) or that pollution control standards be feasible (as in the Clean Water Act), they implicitly embrace a concept of social welfare. What precaution advocates *do* reject is the elevation of maximized welfare over precautionary values. Most also reject modern cost-benefit analysis as a means of measuring social welfare. Cost-benefit skeptics have three main objections to the cost-benefit approach, based on problems they identify with (1) monetization, (2) discounting, and (3) uncertainty.

MONETIZATION

Skeptics argue that the benefits of government protection, whether of human lives and health, of culture, or of natural ecosystems, cannot be meaningfully monetized, because information about an individual's willingness to pay to avoid a risk (or willingness to accept a risk) is unreliable and unduly subject to manipulation. For instance, the wage premium studies on which values of mortality-risk reduction are based assume that workers have accurate information about the risks they face and that they have the power to bargain for the wage premium they think these risks involve. But almost no one believes that these assumptions are true.[38] Plus there is no reason to believe that a worker's willingness to accept a premium for a risk voluntarily assumed at the workplace would be the same as a person's willingness to accept risk involuntarily assumed in the mall, at the park, or at home. In fact, evidence suggests the opposite.[39]

The reliance on wage premium studies also raises environmental justice issues. For instance, studies indicate that minority workers face particular challenges in bargaining for wage premiums to compensate for assumed risk. And the large majority of surveyed workers are male adults, who as a group, some experts believe, are less bothered by risk than women, children, and the elderly.[40] Even if one believes that reliable surveys are possible, there is still the philosophical question of whether the sum of people's willingness to pay can actually be equated with the political will of the citizenry or (even more troublesome) the common good.

Monetizing the value of ecosystem protection is no more reliable. One problem is that understanding the complexity and extent of natural services to the degree of accuracy needed for cost-benefit analysis is extremely challenging and probably beyond our capability. Another problem is that market evidence about an individual's willingness to pay for the consumption or use of a resource does not account for its "existence value": the value one attaches to a natural feature like the Grand Canyon or the Florida Everglades "just because it's there." Survey-based methods, called contingent valuation, try to approximate existence value by asking people how much they would pay to protect places like the Grand Canyon or the Everglades, but this venture is highly problematic and depends a lot on how the question is asked and how much money the responder has.[41] Finally, many people (perhaps most) believe that ecosystems have an inherent value independent of human beings; and this philosophical value cannot be monetized and factored into such a calculus.

For this reason, as I warned in Chapter 2, monetized estimates of ecosystem services should not be swallowed without a pinch of salt. As an illustration, recall the report for the Ramsar Convention on Wetlands that estimated the annual worth of the world's wetlands to total $3,300 per hectare and the annual worth of flood protection services alone to be $450 per hectare. Such price tags might impress the general public and, if they do, may be useful as a way of increasing general interest in environmental protection. But they do not provide reliable guidance for national or global actors. These numbers cannot tell you, for instance, how much a society should be willing to spend to protect a given swath of wetlands. As I mentioned in Chapter 2, the first figure, intended to capture the value of all services, leaves out important services like sediment control and uses an artificial "willingness-to-pay" model to capture cultural value. The calculation omits any existence value that such ecosystems may have that are independent of human interests. The second number, intended only to capture flood protection services, assumes we have reliable knowledge about the extent of services provided. But nearly everyone agrees that we do not have such information and may never know as much as such calculations would require. Thus, a decision-making method based on something other than cost-benefit analysis is needed.

DISCOUNTING

Skeptics of cost-benefit analysis also argue that discounting undermines the value that environmental laws have properly placed on protecting future generations. Cost-benefit advocates favor discounting as a means of estimating the value of a regulation that will impose costs in the short term but will not produce benefits until some time in the long term. Current OMB guidelines recommend that agencies, when conducting cost-benefit analysis, apply a discount rate to any benefit, including preserved ecosystems or saved lives.[42] Thus, applying a discount rate of 5 percent to the death of a billion people 500 years from now would yield the conclusion that such an event is less harmful (or costly) than the death of one person today.[43]

Does it make sense to treat a future saved ecosystem or human life as if it were a future dollar? Cost-benefit advocates say yes. They argue that regulatory benefits should be seen as project outputs you invest in. If the rate of return on a lifesaving project is less than the rate of return on the stock market, then a present generation badly serves a future one when it picks the first, "lower paying" option. It would be better to strengthen tomorrow's economy by using that money to educate more people or create more jobs. Or one could create a financial endowment for future generations and let them spend the more rapidly accruing proceeds to buy benefits (or avoid ills) as they see fit.[44] The point of a free market, after all, is to leave the planet wealthier than when you found it.

Ultimately, this argument fails, for three reasons. First and most obvious, leaving the planet wealthier, in some narrow sense, does not necessarily mean you are leaving it better off. For instance, economist Martin Weitzman argues that some losses, such as the loss of a species or of an ecosystem, are irreplaceable and therefore cannot be compensated for. Serious losses in natural infrastructure can, in fact, lower the future economic growth patterns a present generation can hope to bestow.[45] Imagine a desert city starved for water or a coastal state stunned by the collapse of an important fishery, and you will get the point.

A second difficulty is that when discounting takes place over long periods, the people making the decisions are not the same people who will be living with many of the consequences. As a result, "[n]o one individual will experience both the beginning and the end of the transaction; no one is able to make the personal judgment that the trade-off is, or is not, worthwhile."[46] Even if we assume the future would willingly delegate to the present such important choices, the argument presupposes some formal mechanism by which capital saved through today's foregone projects could be held in trust for future generations, whether in the form of improved technology, infrastructure, or a literal endowment. The logistics involved in developing, maintaining, and policing such a system would be overwhelming—and very expensive.[47] For this reason, no public official that I am aware of has ever proposed such a thing.

Third is the logical problem that if a discount rate is applied for a long enough time, any regulatory benefit, no matter how large, can always be reduced to near-zero. Discounting thus discourages efforts to address problems expected to occur many years from now, like species extinction or incidence of human cancer. Critics have raised this issue in the context of climate change, but it is a problem with disasters in general, since disasters are, almost by definition, singular events with the potential to occur far off in the future.

UNCERTAINTY

In addition to concerns about monetization and discounting, cost-benefit skeptics argue that the lack of information about probability and the magnitude of loss make it nearly impossible to locate an estimated value that is in any way useful. To take one familiar example, the magnitude of cancer risk is notoriously difficult to assess because of the lack of good epidemiological studies on most toxins and uncertainty about how to extrapolate the results of animal-laboratory studies to human beings. Because of uncertainties like these, the range of estimated benefits can vary significantly. Cass Sunstein, a law professor who would later become the Obama administration's regulatory czar, has studied (and defended) cost-benefit approaches for much of his career. After the EPA proposed limiting arsenic in drinking water to 10 parts per billion, Sunstein concluded that with "reasonable assumptions" the rule "can be projected to save as few as 0 lives and as many as 112 total over the years in the U.S." "In these circumstances," he reasoned, "there is no obviously right decision for government agencies to make." Even so, that uncertainty did not shake his belief in the cost-benefit approach as applied to arsenic.[48]

The problem of inadequate information gets worse when scenarios involve overlapping forces that follow complex nonlinear patterns. Weather and climate are notorious for their complexity in this regard. When *risk* is impossible to determine, the decision-maker is left only with uncertainty— uncertainty about the *severity* of harm and about the *probability* of harm. Under such conditions, a monetized value for government protection is almost meaningless, because you don't know how often society will enjoy the benefit of a levee, say, or new earthquake codes—if ever. I'll return to the problem of uncertainty later in the chapter.

Cost-Benefit Analysis and Disaster

For the substantive and procedural reasons just articulated, the cost-benefit project is seriously flawed. Even if one were inclined to overlook some of its shortcomings, it seems unlikely that the "Herculean" costs of this approach justify the insight it is said to deliver. Or to put it another way, cost-benefit analysis fails its own test.[49] Understanding the general weaknesses of cost-benefit analysis will help me show the particular, and more obvious, harm in applying the method to disaster policy. The remainder of this chapter thus focuses on a more limited argument against cost-benefit analysis: that whatever you think of it as it is used in conventional circumstances, it is the *wrong* choice for disasters and extreme catastrophe. When analyzing disasters—from riparian floods to earth-splitting asteroids—a utility principle just doesn't work.

Most of the flaws that exist in the cost-benefit approach regarding monetization, discounting, and uncertainty are grossly amplified in cases of low-probability, high-magnitude harm. Thus, even very dependable cost-benefit advocates now carve out broad space for precaution in cases of catastrophe. The two most notable voices calling for this approach belong to Cass Sunstein and Richard Posner, a federal appellate judge and a senior lecturer in law at the University of Chicago. Disasters are different from other bad events for reasons of speed, scale, and surprise. These characteristics create special difficulties in quantifying and monetizing risk. These difficulties are so vexing that experts like Sunstein and Posner have each devoted entire books to attempts to resolve them. The main problems are as follows.

BIAS TOWARD ZERO

Precaution skeptics suggest that when people are faced with low-probability events, they grow more alarmed than they should be. Before being elevated to the Supreme Court, Stephen Breyer suggested such hand-wringing had led to overregulation in the field of environmental protection.[50] John Graham, the regulatory czar of the second Bush administration, attributed aggressive safety regulations to public "paranoia."[51] Sometimes this is true. People do, in some circumstances, overestimate the risk of very improbable dangers. But that is not the norm. Referring to psychological research, Sunstein writes: "on average, people show *less* concern for [low-probability] events than they should.[52] In general, people are pretty good at comparing, say, a 50 percent risk of a harmful event to a 25 percent risk. But when risks begin to fall below 1 percent, people's eyes glaze over, and they reduce them to near-zero. This is one reason people don't normally think about a major asteroid collision occurring somewhere in the world (with an estimated annual risk of 1 in 1,000) or a serious earthquake striking Manhattan (somewhere around 1 in 40,000).[53] Still, the severity of these events suggests we should have some way of evaluating them. If a magnitude 7 earthquake were to hit Manhattan, for example, half of all structures would receive at least moderate damage, and about 1,700 buildings would collapse completely. Experts say more than 500 people would be killed and another 3,000 hospitalized, and those estimates strike some as dramatically low.[54] Surely some government agency should be thinking about that.

The tendency for people to underestimate low-probability catastrophic risks suggests the need for some policy mechanism to consider them. Sunstein suggests that cost-benefit analysis serves this purpose by making sure that even a 1-in-1,000 chance (or, for that matter, 1 in 1 billion) is calculated into the cost-benefit formula, leading to an "expected value" for the risk (expected value = loss × probability). If a predicted loss is large enough, it will demand attention even if its probability is very low.

Sunstein is right, but only in a limited way. As he concedes, his theory is limited by the problems of uncertainty and what he calls "social amplification." (I will get to those shortly.) But there is a broader problem. The cognitive "bias toward zero" limits our imagination by discouraging us from even thinking about very unlikely events. Because cost-benefit analysis is applied to circumstances already imagined, it cannot push policy-makers to consider the next new "Big One," whatever that might be. Put another way, if most regulators aren't thinking about asteroids or Manhattan earthquakes, what else aren't they thinking about? The seismic explosion of Yellowstone National Park? An accident at a heavy-ion particle accelerator that infects the world with "strange" quarks, crushing the planet into "an inert hyperdense sphere about one hundred meters across"?[55]

Richard Posner's book *Catastrophe* surprised some by dwelling on these and other highly unlikely global disasters. But books like this help to stretch the imagination and urge the public and policy-makers to conceive of the worst. Then a process of evaluation can take place. Literature, particularly speculative fiction, can help in this process. Posner himself was inspired to write his book on catastrophe after reading Margaret Atwood's novel *Oryx and Crake,* a dystopic end-of-the-world story in which most of humanity is wiped out in a perfect storm of bioterrorism, genetic engineering, and climate change.[56] In a law school seminar on disaster law, my students and I end the semester with a discussion of this book. Imagining and then considering the unlikely catastrophe, however, should not be allowed to paralyze decision-makers or distract from greater dangers. A proposed decision-making method must therefore protect against this, too.

SOCIAL AMPLIFICATION

A stronger challenge to a utility principle is that catastrophic events, because of their scale, can amplify resulting harms beyond what a simple body count may suggest. Thus Sunstein suggests that the loss of 200 million people might be more than 100,000 times worse than the loss of 2,000 people.[57] The loss of two-thirds of the country's population could very well threaten the existence of the United States, or at least dramatically change its institutions.

Even smaller, regional disasters can create synergies of doom. Earthquakes, floods, and forest fires do more than kill or injure human beings; they destroy property, interrupt commerce, and damage ecosystems. As we saw in Part II, disasters can also have profound social effects, begetting crime, homelessness, psychological trauma, unemployment, fractured communities, and the loss of cultural heritage. It would be impossible in any meaningful way to assign expected values to such harms for purposes of balancing utility, because these harms are so varied and, in some cases, so far removed from analog markets. (How do you value the loss of your synagogue, the loss of your city?) Thus, the precautionary approach we choose must be able to consider the social amplifications of risk that people experience but that cannot be reasonably captured in market analysis.

RATIONAL RISK AVERSION

Precaution advocates frequently make the point that utility approaches refuse to acknowledge the value of risk aversion. There are some risks that we tend to disfavor regardless of their expressed value. This disfavor is often expressed as fear. Thus, surveys show that the public fears certain hazards, for example terrorism, plane crashes, and exposure to radiation, more than the probabilities of harm suggest they should.[58] Precaution skeptics conclude from such evidence that people overestimate the risk of unlikely threats. (Recall this argument from our discussion of the bias toward zero.) Precaution advocates counter that this amplified fear is really the expression of a reasonable aversion (they call it "dread") to particularly nasty ways of biting the dust, usually because of the accompanying stress, pain, or lack of control. They insist that at least some of these aversions are reasonable and should be considered as part of a political preference for precaution in such areas.

The claim seems true. One reason involves social amplification: people reasonably dread events where social upheaval is part of the anticipated harm. An additional reason has to do with an effect described in prospect theory, a concept economists have used to explain insurance markets. According to prospect theory, people prefer to protect themselves from large losses with low probabilities rather than from small losses with high probabilities, even when the expected value of the small loss is the same or higher. This observation explains why many people are willing to pay insurance premiums that are much higher than the expected values of the losses they are insuring against. (It also explains why a person will spend a dollar for a lottery ticket despite a nearly infinitesimal chance of winning: small losses, even when virtually certain, just don't matter much.)

Some economists believe this behavior is irrational. But as Sunstein explains, it need not be. Precaution in the form of insurance premiums, for instance, may be seen as a hedge against an unlikely scenario that will cause sudden and severe economic loss and create lots of hassle. The "sudden and severe" aspect is important. When a person loses a large investment all at once (like a home) his experienced harm is greater than when he loses a little bit regularly over time. (Which is to say, the last dollar lost in the first scenario will affect his utility much more than the last dollar lost in the second scenario, even assuming the total loss is the same.) The hassle is important because it is a harm that is not likely to show up in a utility-based analysis. Distinguishing *rational* risk aversion in this context from *irrational* risk aversion is complicated, to say the least, and may in the end be in the eye of the beholder. Most of us would agree that insuring one's home against fire makes rational sense for reasons discussed earlier, even if the estimated value of the loss is less than the total premiums paid. In addition to home insurance, Sunstein reports that he pays a premium for emergency automotive service on the grounds that being stranded on the interstate is sufficiently dreadful and inconvenient to require a margin of safety. But he is unwilling to pay to insure his wife's wedding ring because "[i]t's not rational."[59] Sunstein, as his wife might point out, has wandered onto the slippery slope here. But his acceptance of subjective dread, in at least some situations, is significant. Where events like fire and flood are concerned, society's widely shared risk aversion seems not only normal but rational.[60]

UNFAIR DISTRIBUTIONS

Precaution against disaster can also be justified on the grounds that society should pay a premium to avoid unfair distributions of catastrophic harm, even where the likelihood of any particular event is very low. We saw in Part II that disasters are highly correlated with distributional unfairness. One concept of justice, developed by John Rawls, argues that a society should in general provide protection and other benefits on an equal basis but where inequalities arise should cut in favor of "the least-advantaged members."[61] This principle can be used to argue that even if you believe that a precautionary approach should be limited to special circumstances, disasters should qualify as such circumstances because the benefits of avoided harm would inure disproportionately to marginalized groups.

COMPLEXITY AND UNCERTAINTY

Scientists tell us that even with today's mathematics, it is impossible to predict the motions of three pendulums tethered to one another.[62] Many large scale disasters are more complicated even than that. Consider the Dust Bowl storms of the 1930s, as recounted in John Steinbeck's novel *The Grapes of Wrath*.[63] The arid midwestern plains were swept by a series of pendulums—ecological, political, and economic—swinging in an elaborate dance of unsympathetic motion. First came oppressive drought, flaring sun, and crusted soil—the awesome force of nature's unpredictable processes. Collective human forces come into play, creating a feedback loop that intensified the tragedy. In the decades following the drought, the federal government and East Coast banks encouraged farmers to settle and cultivate the Great Plains, which only 50 years earlier had been known as the "Great American Desert."[64] Once established, midwestern farmers were encouraged to grow water-intensive crops, such as corn and Russian wheat, and to engage in "sodbusting." They converted wetlands and tallgrass forests into farms, eradicating thousands of plant and animal species in the process. The sodbusting policy coupled with single-crop agriculture led to the loss of millions of tons of topsoil in the span of a few years. The resulting dust storms lead to thousands of cases of respiratory illness—the "Dust Bowl pneumoni'," as one singer, Woody Guthrie, called it.[65]

Meanwhile, leasehold farmers, desperate to squeeze out short-term profits in a collapsing real estate market, depleted their resources at a rapid rate. Eventually the farms failed, the banks foreclosed, and thousands of refugees set out for California and other western states, where landowners aggressively advertised jobs for migrant workers (even after the jobs were all taken) in an effort to saturate the labor market and drive wages down. What is impressive about the Dust Bowl saga is that almost everything that could go wrong for Dust Bowl farmers did go wrong. In retrospect this series of events may seem inevitable, but if we imagine it as if in the present, we see the forces of history—weather patterns, pioneer migrations, land speculations, sodbusting, widespread illness, and pursuit of jobs in the West—all careening haphazardly. It was as if Steinbeck had drafted his saga with the irregular beats of his contemporary, Igor Stravinsky, playing in the background—which, in fact, he had.[66]

The history of the Dust Bowl is about, among other things, the effect of nonlinear systems: environments where even small changes in one part of the system can produce large and complex effects throughout. Climate and weather patterns are famously nonlinear. Seismic forces are too. Indeed, a wide range of disasters—hurricanes, floods, earthquakes, fires, mud slides, volcanoes, and so on—all march to the rhythm of those three coupled pendulums. What is more, social behavior can also take on nonlinear characteristics (think of crime waves, suburban sprawl, stock market crashes) that can dramatically influence human exposure, vulnerability, and resilience.

The problem with complexity is that it makes estimating the probability and magnitude of harm impossible as a practical matter. Without a reasonable range of probability or magnitude, there can be no expected value (which is the product of those variables). Without an expected value, there is no *risk*. Because the benefit in cost-benefit analysis is "avoided risk," there is no way to estimate the benefit and so nothing to compare the cost to. We are left with a number-hungry cost-benefit machine cast adrift in an inky void called uncertainty.

Cost-benefit advocates have tried mightily to paddle back into familiar waters. They sometimes point to sophisticated statistical procedures intended to make sense of complex systems, including the dark machinations of climate change. So-called Delphi analysis gathers subjective assessments of unknown risks from experts and applies probability theory to find points of convergence. Another powerful method is the "Monte Carlo" technique, in which computers generate thousands of hypothetical distributions of unknown probabilities "with the goal of locating policy prescriptions that predominate over a wide range of possible conditions."[67] Such analyses may be useful in generating possible scenarios and stretching the policy-makers' imaginations, but they are not good at setting meaningful boundaries or separating the things we should worry about from the things we shouldn't. Whatever their value in other endeavors, these techniques are unlikely to produce the reliable, fixed-range data points that cost-benefit analysis demands.[68]

IRREVERSIBILITY

Environmentalists often justify precaution on the grounds that the harm to be avoided is irreversible. The sardine fisheries in Monterey, California, will never rebound from their collapse. Now that we have allowed channelization, inflexible levees, and greenhouse gases to sink the Louisiana marsh, we will never get it all back again. The same logic applies to culture, heritage, and communities of historical interest. These things are often more relevant in disasters. It's impossible to protect miles of forests from being buried by an exploding volcano (as residents in the Mount St. Helens region learned years ago). But society can often reduce natural hazards that threaten vital communities and features of historical importance.

Disasters are an obvious source of irreversible damage, both natural and cultural. Why take special care to avoid irreversible damage, as opposed to the reversible kind? There are two answers. The first is best understood in terms of what economists call an option value. Imagine that most Americans have no current use for an obscure species of spider or a windswept tallgrass prairie. Or imagine that most Americans lack an appreciation for the hunting and fishing traditions of the Biloxi-Chitimacha Indians or the schoolhouse of street jazz once called the Lower Ninth Ward. A failure to protect any of those things might not have much effect on the sum of all American welfare. But by extinguishing the option of having such things in the future, we run the risk that (1) new knowledge about or appreciation for what is lost will cause us to regret the damage in the future, and (2) future generations who have no current say in the matter will be deprived of something they could have put to great use (again because of new knowledge or appreciation). All things considered, it is better to preserve options for ourselves and our progeny than to let them expire. This view does not mean that present generations should never use or modify cultural or environmental resources. But when given the choice of using these resources in a way that preserves future options or in a way that destroys them, we should choose the former. There was a time, remember, when American pioneers judged Native American culture as "savage," and the Grand Canyon a very inconvenient hole in the ground.

The second reason for avoiding irreversible harm is an ethical one. Many people believe that species, ecosystems, and even geological features have some entitlement to our stewardship that is independent from human utility.[69] An easier case is that the human communities, traditions, and heritage that are put at risk by regional disaster also have an inherent entitlement to exist that cannot be traded away simply because the majority does not adequately value it. Indeed, the recognition of the inherent value of culture and heritage can be found in both American and international law. This justification, based on a group-based substantive entitlement, may also be framed as an interest in fair distribution, as I have shown already in environmental and disaster justice arguments (Chapters 6 and 7).

When you stop thinking about morality and group interests, the mathematical mind can play terrible tricks. Take, for instance, Posner's analysis of world annihilation in which he decides, by way of a perverse "guesstimate," that the survival of the human race is worth $600 trillion.[70] I say "guesstimate" because, as he admits, the number depends on not a single verifiable data point. I say "perverse" because the implied message is that if human survival *did* cost more than $600 trillion, we would all be better off dumping our money in an underground vault and waving goodbye. To be fair, I don't believe that a theory's most extreme conclusion necessarily disqualifies it from use in more moderate circumstances. (The precautionary principle, incidentally, is entitled to the same courtesy.) But where we are talking about extreme events, we at least need to understand how far a proposed method might take us.

The foregoing arguments do mean that irreversible harm must always be avoided. Sometimes that mission is impossible or so grossly burdensome that it should not be attempted. Sometimes, as Sunstein has pointed out, there is irreversibility on all sides. There is plausible evidence to suggest, for instance, that genetically modified crops promise (1) to bring food security to sub-Saharan African, and (2) to dismantle the agrarian culture of Mexico. Playing the irreversibility card does not win the game. But irreversibility, particularly in the cases of low probability and severe harm, must clearly be given independent significance in policy analysis. Because the cost-benefit approach does not do this, it is inadequate for disaster scenarios.

Mend It or End It?

Given the shortcomings of cost-benefit analysis when applied to low-probability, high-magnitude events, one inevitably must ask whether, under such circumstances, the approach should be mended or ended. Sunstein and Posner both vote for mending, but ultimately their arguments run out of steam. To understand their reasoning, it is helpful to divide catastrophes into events for which some reasonable attempt at estimating expected value is possible (call this a context of "quantifiable risk") and events for which it is impossible (a context of "nonquantifiable uncertainty"). Sunstein proposes fixes for both situations. Posner focuses mainly on the second.

QUANTIFIABLE RISK

Where estimating expected value is possible, Sunstein suggests that cost-benefit reports be supplemented with information about collective harm—what he calls "social amplification"—to capture a more accurate scope of the harm. He would also apply a "margin of safety," based on "cost-effectiveness," as a kind of social insurance to protect people from dreadful events for which they have a rational risk aversion.[71] Sunstein doesn't say what form the social amplification information would take, but we might imagine a vivid narrative description of, say, sunken neighborhoods, city-wide trauma, and dismantled government. Sunstein's margin of safety is similarly vague; it is not even clear whether the concept would be expressed as a percentage or as a narrative directive. Sunstein would also include a modest discount rate to be applied to the monetized estimated values.

Does this help? The admittedly flawed process of monetization and discounting is still there. The expected value, based on unreliable and biased wage premium studies, still distorts the analysis. The bias in favor of current generations at the expense of future generations is preserved. That both of these biases are hidden from the public beneath an obscure formula of statistical analysis should also bother us. Some of these objections are not particular to catastrophic events, although the concern for future generations is, if only because disasters tend to spread over longer time lines. Suppose we are looking at a Katrina-like event that has the potential to kill a thousand people but which will probably not occur for another 100 years. A modest discount rate of 3 percent would shrink the human loss to a present value of 52 lives, or (using the EPA's valuation) about $317 million. But that small number seems embarrassingly low. If Sunstein's margin of safety is an actual percentage, one might imagine it could help drive the monetized benefit upward (in the name of avoiding dread and hassle) but logically that margin should be discounted, too, as its advantages are experienced deeper into the future.

The narrative supplement—which will include a description of social upheaval and might include a nonquantified margin of safety—is also problematic. The monetized ratio of cost to benefit seems intended to set a presumption in favor of the higher value. If monetized costs outweigh monetized benefits, how often would a disaster narrative or margin of safety be permitted to override that conclusion? The question is crucial, but neither Sunstein nor any other cost-benefit advocate I know gives it serious attention. Hard numbers, because they are easy for the brain to process, may persuade people more easily than abstractions.[72] No matter how large the uncertainties behind the numbers, one suspects that a disappointing net benefit range of negative $50 million to positive $1 will almost always trump "avoided social trauma" or "enhanced distributional fairness." When Sunstein has previously recommended narrative supplements to the cost-benefit ratio (to take into account things like distributional unfairness), he has insisted that such supplements would override the ratio only in exceptional cases.[73] That is hardly a way to correct for a flaw that Sunstein argues is pervasive in the evaluation of catastrophic events. On the other hand, if he allows the narrative to swallow the ratio, he is left to admit that all the fancy number crunching is really not the most important part of disaster mitigation.

NONQUANTIFIABLE UNCERTAINTY

In the face of what Posner calls "radical, nonquantifiable uncertainty," cost-benefit advocates have a special challenge, since they have no estimated value for risk. Posner and Sunstein offer different solutions to the problem. Their best proposals, though, look suspiciously similar to the precautionary approach embraced by environmentalists. Posner offers at least three possible solutions to the problem. One is what he calls "inverse cost-benefit analysis." Under this proposal, one divides "what the government is spending to prevent a particular catastrophic risk from materializing by what the social cost of the catastrophe would be if it did materialize."[74] Thus, if you want to know the probability implied in our expenditures on a Pacific Coast tsunami, you look at what society is spending on that problem and divide by the monetized expected loss (probability = expenditure ÷ loss). Here one assumes that society is already using cost-benefit analysis (at least in an unconscious way) so that its expenditure is the same as the expected value of the risk.

The trouble, of course, is that there is no evidence that society is weighing costs and benefits at all in this situation, so the meaning one can attach to social expenditure is exceedingly slim. Leaving that argument aside, the most you get from this method is society's *subjective* view of an event's probability, which everyone would agree is ill informed. (It has to be ill informed: the probability is statistically nonquantifiable, remember?) The *objective* probability of a Pacific Coast tsunami remains unknown, and it is that information that lives depend on. For these reasons, even Sunstein rejects Posner's inverse cost-benefit approach.

A second alternative offered by Posner is what he calls the "tolerable windows" approach. He writes:

> Suppose the optimum cannot be determined because of uncertainty about costs, benefits, the discount rate, or probabilities. We may nevertheless know enough about the benefits and costs to be able to create [a] "window" formed by the two vertical lines. At the left side of the window frame the benefits of a further effort to eliminate or prevent the catastrophe in question comfortably exceed the costs, while at the right side the reverse is true. If we stay within the window, although we won't know whether our measures are optimal we'll at least have some basis for confidence that they are neither grossly inadequate nor grossly excessive.[75]

The approach backfires, though, because it is based on exactly the kinds of information (cost, benefits, discount rates, probabilities) Posner says we do not have. Even assuming that some minimum amount of information existed, the method does not seem helpful in circumstances as in some kinds of natural catastrophe, where the size of an event increases much more rapidly than the frequency. (Earthquakes are a classic example: an earthquake twice as large as another is four times as rare.) In such circumstances (which experts call a "power law" relationship), the window of action would be exceedingly wide, making too many options available and defeating the purpose of the exercise.

Posner suggests a third alternative while discussing defense against asteroids, although it would seem to apply to any extreme event where probabilities are unknown and the benefits are not monetized. In such circumstances, he says, the government should provide information about the costs of protecting society against catastrophe and the death rates that can be expected from a variety of different outcomes. "The government and the public would then have the information required for rationally deciding whether the nonmonetized benefits of the statistical lives saved . . . were worth the costs of saving them."[76] This flexible approach avoids the main criticisms of cost-benefit analysis and is much more transparent. In fact, it resembles "open-ended balancing," the precautionary method used in licensing pesticides. Lisa Heinzerling has argued that Posner's last alternative is just another version of the "holistic" analysis that she and economist Frank Ackerman have developed as a substitute for the cost-benefit approach.[77] For this reason, Heinzerling calls Posner an "accidental environmentalist," a theorist who despite his best efforts has stumbled into the tree huggers' camp. Environmentalist or not, Posner's idea still needs fleshing out. It does not, for instance, explicitly consider social amplification, unfair distribution, rational risk aversion, or irreversible environmental damage (although decision-makers could, if they wished, consider those things). The approach may also be susceptible to the "bias toward zero," depending on which scenarios are chosen for examination (if the extreme ends are presented as outliers, they might be dismissed). In addition, the approach provides no guidance for how the information would be organized for review by policymakers or members of the public.

Now let's look at Sunstein's proposal. His approach to nonquantifiable uncertainty begins with a decision rule used in game theory called the maximin. The rule, which is also associated with Rawlsian philosophy, holds that when choosing among options, a decision-maker should minimize the maximum loss, that is, choose the "least bad outcome." Sunstein knows the maximin can lead to paralysis if taken to the extreme. We would never build a dam or even cross the street if we sought always to avoid the worst imaginable case. In addition, *not* building the dam or crossing the street might produce its own tragic outcome. At this point, the rule becomes incoherent, advocating inaction and action at the same time. (It's true that the bad consequences of not building the dam might be worse than the bad consequences of building the dam, but in a context of uncertainty, how do you know?) Sunstein fine-tunes the maximin rule by factoring in cost-effectiveness (which he says Rawls also intended) and risk-risk trade-offs. I cannot describe Sunstein's multipronged rule more succinctly than he already has:

> In deciding whether to eliminate the worst-case scenario under circumstances of uncertainty, regulators should consider the losses imposed by eliminating the scenario, and the size of the difference between the worst-case [scenario] under one course of ac-

tion and the worst-case scenario under alternative courses of action. If the worst-case scenario under another course of action is much worse than the worst-case scenario under another course of action, and if it is not extraordinarily burdensome to take the course of action that eliminates the worst-case scenario, regulators should take that course of action. But if the worst-case scenario under one course of action is not much worse than the worst-case scenario under another course of action, and if it is extraordinarily burdensome to take the course of action that eliminates the worst-case scenario, regulators should not take that course of action.[78]

After a careful reading (or perhaps several) what you get is a version of the precautionary principle. While it considers cost ("if it is not extraordinarily burdensome"), it does so only in relation to what society can bear, not in relation to the putative benefit. In this way, it follows the injunction of the feasibility principle: do the best you can. Sunstein's approach could be criticized for assuming, despite posited uncertainty, that alternate worst cases can be compared. ("If the worst-case scenario under another course of action is much worse than the worst-case scenario under another course of action.") But the gist is definitely precautionary; it may be that failings like these in extreme situations simply have to be borne.

It is important to point out that both Posner and Sunstein assume that examples of nonquantifiable uncertainty will be rare. In his book on worst-case scenarios, Sunstein's maximin serves as an emergency parachute only when cost-benefit analysis falls into a death spin. Similarly, Posner reserves his precautionary theory for the near-apocalypse, situations in which the fate of humanity is literally at stake. However, as Daniel Farber and others have suggested, nonquantifiable uncertainties may be more common than we think.[79] This certainly seems true where weather and climate are concerned. Therefore, Sunstein's and Posner's proposals offered as last straws may have application in a wide range of disaster scenarios. In addition, recall that even in cases where probability is quantifiable, Sunstein's approach to disasters (narrative supplements and margins of safety) does not solve the problems he identifies. This suggests the need for an even broader application of a worst-case decision method.

We are getting closer to a workable approach. In Chapter 10, in fact, I will describe a precautionary strategy for disaster analysis that draws from Posner and Sunstein. Before that, however, it is important to see the abstract in a more concrete way. Without actual context, it is hard to see why nonquan-tifiable uncertainty is more prevalent than we think, or how malleable and obfuscating hard numbers can be. We should also keep in mind that good decision-making usually takes more time, leading to a delay in action. The delay may be worth it, but it should be examined. To put flesh on the bone, let's consider two more stories from New Orleans.

9

MAPPING KATRINA

This chapter offers two stories about risk management to illustrate some of the points made in Chapter 8. The first story looks at the government's development of the protection standards used in protecting New Orleans from storm surge. The process was based on an antiquated version of the cost-benefit approach, intended to maximize aggregate social welfare. It also relied on something called a "project hurricane," a fictional event made up on the basis of a composite of storms that had occurred in the past. There is plenty to criticize in this approach—especially given its tragic failure. But the take-home message has much broader application: in the context of deep uncertainty, mathematical models designed to pinpoint project benefits by focusing on a single worst case will invite failure; and when the models fail, they can fail spectacularly. Taken together, the theory and the example suggest that disaster analysis must move away from a utilitarian approach and embrace a precautionary model instead, one that has the ability to evaluate several possible worst-case scenarios rather than just one.

The second account lesson comes from a more specific aspect of the Army Corps's storm protection project: its abandonment of a plan to install barrier gates after its environmental review of that plan was rejected by a federal court. After Katrina, some claimed that the requirement of an environmental review pushed the Corps toward a less protective option. Full analysis will reveal that this precautionary approach did not contribute to the New Orleans flood, after all. The lesson here is that precautionary analysis, properly employed, allows decision-makers broad flexibility to protect public safety; so the examination of alternative plans in the name of precaution should not be feared.

Imagining the Perfect Storm

In 1965, Hurricane Betsy severely flooded the poorer eastern part of New Orleans, as well as neighboring St. Bernard Parish.[1] While the damage was nowhere near that of Katrina, local and federal agencies urgently sought a plan to protect the Crescent City from future storms. Officials at the Army Corps of Engineers rolled up their sleeves, unsheathed their slide rules, and went to work. The first step was to determine what kind of storm a future system should protect against. When the Netherlands was crippled by a North Sea storm in 1953, the Dutch resolved to build a system of dikes that would in some areas protect against a 10,000-year storm, that is, a storm with a 0.01 percent chance of occurring in a given year. The Corps chose a lesser goal; it adopted a hypothetical hurricane, called the Standard Project Hurricane, that it would design a levee system to protect against. The National Hurricane Research Program had given this model storm various attributes, all built around the predicted central barometric pressure of a 100-year storm, a storm with a 1 percent chance of occurring in a given year. As Douglas Kysar has noted, there is no satisfactory answer to the question why government officials focused on 100 years as opposed to 1,000 or 10,000.[2] Commentators at the time suggested the model described the worst storm that was "economically reasonable" to protect against. Indeed, some had argued that even that level of protection was overly cautious. In reviewing the history of the Standard Project Hurricane, a 1972 government update suggests that the choice of 100 years was simply "arbitrary."[3] What we can say is that the choice was not solely a technical one. Indeed, years later the Standard Project Hurricane was revised to reflect harsher possible storms.[4]

By these questionable means, then, the safety standard (a "100-year storm") was set. But there was also a problem in how it was described. Imagining the characteristics of a 100-year storm requires studying years' worth of data on previous storms in the region. But in the 1950s, the time of this modeling, scientists confronted two serious limitations. First, meteorologists had very limited empirical information, what journalists John McQuaid and Mark Schleifstein have called "a Swiss cheese of statistics and meteorological knowledge" drawn mainly from regional storms occurring between 1915 and 1947.[5] Second, because much of the data were extrapolated from on-shore measurements (before the use of aircraft to read storm pressure) the records that did exist were considered unreliable. In addition to these obstacles, it is also possible that the scientists themselves contributed to inaccuracies by dropping two particularly severe hurricanes—Camille and the "Keys Storm," which they believed were "not reasonably characteristic" of the Gulf region.

As risk management, this identification and design for the standard project hurricane suggests two flaws, both related to spatial geography. The first flaw was the decision to provide only "100-year" protection to a city whose history and geography clearly demanded more. At the time New Orleans was already more than 200 *years old* and showed no signs of serious decline. The Gulf Coast had already witnessed two much stronger hurricanes less than fifty years earlier (Camille and the Keys Storm, both later dropped from the storm modeling, as noted). In addition, the city housed some of the most cherished culture and architecture in the United States. Second, the 100-year standard was chosen by unelected agency officials with little or no input from the people whose geographic areas would be affected.

But the largest concern raised by the story is the failure to appreciate the uncertain nature of the *temporal* landscape. Weather and climate are notoriously difficult to read, given the countless factors involved and the global scope of weather dynamics. In addition, storm force depends on geographic features that change over time, including land subsidence, coastal erosion, channeling, land-use development, and ocean temperature. In complex systems such as storm cycles, normal probability distributions are not always, well, the norm. Extreme events can seemingly pop out of nowhere or arrive together in swarms. Remember that the 2005 Atlantic hurricane season saw a record 15 hurricanes, a record 27 named storms, and the most Category 5 hurricanes recorded in a single season (Katrina, Rita, and Wilma).

The implementation of more than a hundred miles of levees was strongly controlled by cost-benefit analysis. The formula used was extremely narrow by modern standards. According to the Corps's guidelines, flood control projects were measured in terms of "net economic development" (NED). The system required the Corps to justify its projects only in terms of new development, which for levee projects usually meant recovered land in floodplains that could be opened for residential or commercial development.[6] The benefits of protecting *existing* buildings or infrastructure did not count, nor did other obvious benefits like saved lives or avoided injuries. This system, which has been in place for decades, creates the richest of ironies: the most easily justified flood control projects in the Gulf are those that actually *encourage* relocation to undeveloped marshes and swamps.

The Corps later developed a way of referencing environmental assets protected by proposed projects, but this innovation has never competed with NED on anything near level playing field.[7] The Corps also required the local levee boards to contribute up to 30 percent of project costs as a way of confirming local commitment to the project. Finally, the Corps's "factor of safety" for levee design is an incredibly modest 1.3, meaning that the structure needs to be designed to withstand only 1.3 times the predicted force. That leaves little margin for error. Had the Corps been required to use the much higher construction standards required for bridges or dams, it is likely that New Orleans would have remained dry.[8]

The second flaw concerns the NED guidelines. The narrow focus on new development follows from a general rule that levee projects at the time were rural projects far away from large populations and existing structures. This assumption also explains the low factor of safety in levee design. Why knock yourself out to protect pastures and cattle? The problem is that not all geography corresponds to general assumptions. The levees in southern Louisiana protected a metropolitan area of more than 2 million people. Geographic characteristics also change over time. In California's San Joaquin Valley, to use an example from another state, miles of modest levees that once protected rural farmland now protect a metropolitan area of hundreds of thousands of people.

The NED's reliance on the development value of a newly recovered swamp is an indirect way of inquiring into how much the public would value a levee that made such development possible. The slippage in this process is enormous, since the many other public values associated with levees, such as saved lives, avoided injuries, and intact communities are not reflected in development values. The slippage gets worse when one considers the Corps's requirement of a 30 percent contribution from the host state. For even if the Corps determines that a project is justified economically on the basis of only new development, it will still languish if the host state refuses to contribute that 30 percent. The requirement, in effect, adds *another* willingness-to-pay inquiry, directed this time to only *local* citizens. Is this objectionable? In the case of southern Louisiana, I think it is.

The problem is that this local willingness-to-pay estimate is kept low because of two reasons related to place. First, Louisiana is a relatively poor state. It routinely scores near the bottom of national assessments on crime control, employment, and economic growth. The causes run deep and are themselves related to geography, including the vestiges of slavery in the agrarian South, the failure of "Reconstruction" in the late nineteenth century, and a more modern federal energy policy that allows the gas and petroleum industries to decimate the coast while at the same time denying Louisiana a fair share of the compensatory royalties collected by the federal government. Louisiana's relative poverty is important, because its modest "ability to pay" necessarily depresses its willingness-to-pay amount. Where the willingness-to-pay finding is used as a proxy for "the public will," a regulator is likely to draw the false conclusion that wealthier states value storm protection more than poorer states because they are willing to pay more for it.

Second, local actors, particularly less affluent ones, are susceptible to a form of consumer discounting in which they will prefer smaller, earlier benefits (including cost avoidance) to larger, later benefits when the smaller benefits are imminent. In the current context, Louisianans are susceptible to discounting because as local taxpayers they are able to capture almost immediately the benefit (in the form of saved costs) of not investing in levees. They can therefore be expected to discount the benefit of levee protection more deeply than if the saved costs were captured less imminently. The fact that the state of Louisiana is poorer than most other states in the United States only exaggerates this effect, because the marginal utility of the saved costs is that much greater. In this way, Louisianans resemble hard-luck fishers in a declining fishery. Even though they know aggressive catch limits would promote future sustainable harvests, they prefer more permissive catch limits now because they perceive their immediate harvest to be much more valuable than an even bigger harvest tomorrow (or, perhaps, *any* harvest tomorrow). In contrast, we can expect the public at large, which also has an interest in fish productivity but does not profit as directly from the present harvest, to be more likely to prefer imposing aggressive catch limits now.

Did Precaution Sink New Orleans?

Less than a month after I evacuated from New Orleans and temporarily relocated to Houston, I was asked to testify before a congressional subcommittee on the environmental effects of Katrina. Parts of New Orleans were still underwater. Many environmentalists had begun criticizing the oil and gas industries for eroding the wetlands and the chemical storage facilities for failing to prevent toxic spills. But in Washington, a group of conservative commentators had begun to turn the tables. They suggested that the New Orleans Flood was caused not by too little environmental precaution but by too much. Specifically, they argued that a 1970s-era environmental dispute over an environmental impact statement (EIS) had prevented the Army Corps from building stronger storm barriers. Their reasoning was wrong, but the story packed punch, and it gradually made the rounds on Capitol Hill.

I was reluctant to travel to Washington, then, because so many other concerns had taken over my life. My kids were enrolling in new schools. My house, having stood in five feet of water for three weeks, was becoming a mold farm. My insurance agent wouldn't return my calls. And I didn't have a suit. Still, a Republican staffer assured me that my personal perspective would be valuable. I agreed and was soon on a plane to Washington, tapping in last-minute revisions to my testimony along the way. I entered the Cannon Office Building, wearing a new jacket and slacks and tugging a rolling suitcase. The staffer I had spoken to earlier found me in the hearing room and introduced himself. He looked much younger than I had imagined, about the same age as one of my law students. He smiled broadly, and I began to relax. "Hey, professor," he said, "looks like your environmentalist friends just killed a thousand people."

During my testimony, after I had spoken about eroded wetlands and toxic spills, I was, of course, asked about those abandoned storm barriers. I would later answer the same questions during testimony before a Senate committee as well. The questions vexed me at the time, because they seemed insincere, delivered with an almost secret delight. Still, the concern *is* valid and deserves analysis. In Chapter 8, I praised the precautionary approach that is used in environmental law and argued for its extension in disaster mitigation. In response to the objection that precaution in one direction can expose you to dangers from another, I minimized the concern, saying that decision-makers were likely to focus on the most threatening hazard. Yet here is a claim that the crown jewel of our nation's precautionary efforts, the EIS, may have led to the flooding of a beloved American city. Is there any truth to this claim? In the future, how can we minimize the chance that we will trade in one bad outcome for an even worse one? Let's take a closer look at this controversy.

The argument that environmental review sank New Orleans arose during the immediate political aftermath of the New Orleans Flood. While people were still huddled on rooftops awaiting rescue helicopters, politicians and pundits began pointing fingers in what President Bush would later call the "blame game." Amid this debate came a claim that New Orleans's levee system would have protected the city had it not been for a lawsuit filed in the 1970s by local fishers and a local environmental group under the National Environmental Policy Act (NEPA).[9]

Enacted in 1970, NEPA requires all federal agencies to prepare EISs for all "major federal actions significantly affecting the human environment."

A trailblazer at the time, NEPA's precautionary "look before you leap" mentality became the foundation of environmental policy in the United States and around the world. Its directive to consider potential environmental impacts and feasible government alternatives has "unquestionably improved the quality of federal agency decision-making" in environmental terms, minimizing the risks of dangerous emissions, saving billions of dollars in inefficient construction plans, and helping to provide essential habitat for dwindling species.[10]

Adherents of the theory that NEPA was to blame for the levee failures argued that when a federal district court blocked work on the Army Corps's original levee plans, pending an adequate EIS, the Corps was prevented from building a project that would have protected the city from Katrina's devastation. Instead, this argument goes, the Corps chose a second, inferior design that ultimately could not protect the city.

Here's the full and true story. In the late 1960s, after Hurricane Betsy had pummeled New Orleans, Congress approved a massive hurricane barrier project to shield New Orleans from storm surge. The project, which, was called the Lake Pontchartrain and Vicinity Hurricane Project, was intended to protect the city from possible surges on Lake Pontchartrain to the north, on Lake Borgne to the east, and on swamp waters from the southeast. To realize the goal, the Corps carefully considered two options, one called the "high-level" option and the other called the "barrier" option.

The high-level option required raising many existing levees around the city and building new levees where needed. This option was designed to protect against a hypothetical project hurricane that was loosely equivalent to a fast-moving Category 3 storm on the Saffir-Simpson scale (although at the time that scale did not exist). Under the high-level option, levees would have been raised from 9.3–13.5 feet to 16–18.5 feet above sea level. Needless to say, plans for this option assumed all components would be properly designed and constructed so as to meet these expectations.

The barrier option was also designed around the project hurricane but adopted a different theory. This option called for building a series of levee walls to prevent storm surge from barreling into Lake Pontchartrain from Lake Borgne and the surrounding wetlands. By keeping storm surge from entering Lake Pontchartrain, the city's northern levees and canal floodwalls would be protected from overtopping and extreme pressure, thus obviating the need for building them higher. The barrier would have started at a point near the Mississippi state line and run southwest across miles of marshland to meet up with existing levees near the Gulf Intercoastal Waterway. Where Gulf waters enter Lake Pontchartrain through a "bottleneck" of narrow channels (the Rigolets and the Chef Menteur passes), the Corps would have built a pair of massive sea gates designed to close during times of extreme weather. Gates like these are, in fact, used to protect the Netherlands from storm surge on the North Sea.

The Corps chose the barrier project. The high-level option, it reasoned, presented too many difficulties. For one thing, the high-level option would have required the agency to obtain rights-of-way to widen and raise existing levees, upsetting neighbors and raising costs.[11] For another, it would have left some industrial areas and certain sections of developable wetlands open to flooding.[12] One advantage of the barrier option, in contrast, was that it would shield these open areas and allow residential development to take place, thus increasing the value of that land.[13] Finally, my conversations with past Corps officials suggest that many Corps engineers simply believed that the barrier project would be more effective in protecting against most kinds of storms.

Construction began on the project in the early 1970s but soon encountered delay. Owners of land needed for the new structures demanded high prices for their property and forced the Corps to bring condemnation actions. In 1976 a coalition of local fishers and an environmental group called Save Our Wetlands sued the Corps, arguing that it had submitted an inadequate EIS in violation of NEPA. The plaintiffs had many concerns, among them a fear that the gates would impede natural water circulation and damage the sea life. They were also concerned that residents living outside the protective structure would be prone to much greater flooding when surge ricocheted off of the walls and toward their neighborhoods. It was no help to the Corps that its final EIS, representing the entire environmental review of a multiyear, multimillion-dollar project, consisted of only four typewritten pages. In 1977, a federal district court agreed with the plaintiffs, expressing concern that the Corps had not given adequate consideration to environmental concerns, project alternatives, or costs and benefits. The court issued an injunction preventing further progress on the barrier option until the Corps revised its EIS to correct these errors. Five years later, having never completed a revised EIS, the Corps abandoned the barrier option and adopted the high-level option instead. It was this plan's protection system that in 2005 stood between New Orleans and Hurricane Katrina's muscular surge. And it was this network of levees, as we know, that cracked, buckled, and collapsed in more than a 50 different places.

Less than two weeks after the 2005 flood, debate over the *Save Our Wetlands* litigation and the barrier plan resurfaced in a front-page story in the *Los Angeles Times* "A Barrier That Could Have Been."[14] The story claimed that the barrier project had been "derailed" and "stopped in its tracks" by the environmental lawsuit and questioned whether that litigation might have doomed New Orleans to a lesser flood control system. Experts featured in the article appeared divided on the issue, but an opening quotation from John Towers, a retired chief counsel of the Corps's New Orleans District, clearly set the tone, insisting: "if we had built the barriers, New Orleans would not be flooded."[15] Later in the article, Towers, who oversaw the *Save Our Wetlands* litigation on behalf of the Corps, further lamented the court's decision in that case. "My feeling was that saving human lives was more important than saving a percentage of shrimp and crab," he said. "I told my staff at the time that this judge had condemned the city."[16]

After that, the fur flew. Drawing from Tower's words, conservative pundit R. Emmett Tyrell, Jr., called the plaintiffs "environmental fanatics with no sense of a broad-based commonweal" and argued that NEPA had given them a "veto over . . . projects essential to the health and well-being of millions of Americans."[17] A right-wing blogger branded the levee failure a "Green Genocide."[18] At the request of Senator James Inhofe, the Justice Department opened an inquiry into any other environmental challenges that might have involved the New Orleans levees.[19] And a House task force, already assigned to review NEPA for possible overhaul, added the *Save Our Wetlands* litigation to its list of possible justifications for amending the statute.[20]

It is unfair and destructive to cast responsibility for the failure of the New Orleans levee system on this small band of activists and a popular environmental law. The causal relationship is virtually nil; and the suggestion that weaker NEPA requirements would better protect people from disasters is dangerous. Stronger environmental requirements, if focused on natural services, lead to *more* protection, not less. To understand the charge's unfairness, we must divide it into its two implied assertions, one related to law, the other to engineering. The first assertion is that the Corps's loss in the *Save Our Wetlands* litigation precluded the barrier option, forcing the Corps to resort to its less preferred high level option. The second assertion is that had the Corps built the barrier option instead of the high level option, the New

Orleans flood would have been minimized, or perhaps even avoided. Neither assertion is supported by the facts.

The argument that the judge's ruling in *Save Our Wetlands* precluded the barrier plan does not hold because that is not what the ruling says. To begin with, NEPA does not require that any federal action be implemented or abandoned. The statute requires only that federal actions covered by the statute be subjected to environmental review, which in this case meant a proper EIS. Further, an EIS is a comprehensive document. Today EISs are hundreds, even thousands of pages long; but even in the 1970s, when NEPA was still being defined by the courts, federal agencies knew that courts wanted a serious effort. In addition to length, there were several reasons to doubt the seriousness of the Corps's effort. The court found, for instance, that optimistic conclusions about the effects on sea life were based on an outdated study modeled around an obsolete version of the project. (The Corps's environmental staff had apparently requested new studies, but they had not been completed in time.) The court also expressed concern over the Corps's biological analysis, which "relied entirely on a single telephone conversation with a marine biologist who was asked to speculate about the impact of the project on marine organisms using the inter-lake flow rates predicted by the obsolete model."[21]

In addition, as government lawyers know, NEPA allows unlimited do-overs. An agency can always come back with an improved EIS to justify its proposal. The court even reminded the litigants that its opinion "should in no way be construed as precluding the Lake Pontchartrain project as proposed or reflecting on its advisability in any manner."[22] In fact, the court wrote that "upon proper compliance with the law with regard to the impact statement this injunction will be dissolved and any hurricane plan thus properly presented *will be allowed to proceed*."[23] This judge was not, as the *Los Angeles Times* reported, stopping the Corps in its tracks. He was saying, "Ya'll come back."

The Corps did not come back; instead it dropped the barrier option and took up the high-level option. The explanation for this action involves many factors and is still open to some speculation. In the months after the storm, two law professors, Thomas McGarity and Douglas Kysar, put the process under the microscope and settled on two time-honored explanations: money and politics. The politics became evident when, after the injunction, an "intense public opposition" arose based on concerns for the environment, fears of increased flooding for properties outside the barrier walls, and the resentment from citizens who saw the project as a "land grab" intended to benefit local developers.[24]

At the same time, the project was getting more expensive because of new design requirements and general delay, a consequence of the public opposition as well as physical "foundation problems discovered after project initiation."[25] The barrier option, which in 1965 was estimated to cost around $85 million, by 1982 had become a nearly $1 billion project.[26] Originally, the high-level option had been viewed as the more expensive one, which is why, according to the U.S. Government Accountability Office, the Corps originally discarded it.[27] But by 1981, the updated estimates of the barrier option had surpassed the updated estimates for the high-level option by about 20 percent.[28] Thus, the Corps officially switched positions and selected the high-level option because, in the words of the Government Accountability Office, "it would cost less than the barrier plan" and because it would "have fewer detrimental effects on Lake Pontchartrain's environment."[29]

One can say that environmental concerns contributed to the political opposition and the litigation provided the information and time for activists to organize. But at the same time, one must acknowledge that nothing prevented the Corps from acting more quickly to perform a sufficient environmental review and get the project approved. Had it believed that superior benefits of the option justified increased costs, it could surely have moved forward.

And there is a second broken link in the causation chain: according to several engineering reports, catastrophic flooding would almost certainly have occurred even with the barrier plan. As an official of the Government Accountability Office stated in his congressional testimony, a month after the storm, "[n]one of the changes made to the project . . . are believed to have had any role in the levee breaches recently experienced as the [high-level option] was expected to provide the same level of protection [as the barrier option]."[30] In fact, because of the direction of Katrina's surge, he noted: "Corps officials believe that flooding would have been worse if the original [barrier option] design had been adopted."[31] At any rate, the Corps does not appear to have viewed the barrier option as significantly more protective when it originally chose that option in 1965, since as the Government Accountability Office found, the primary reason for discarding the high-level plan had to do with cost.[32] This analysis finds support in engineering reports later issued by the American Society of Civil Engineers and by

Team Louisiana. Both reports find that the hurricane's surge entered eastern New Orleans, including the Lower Ninth Ward, from Lake Borgne and "Mr. Go," *not* from Lake Pontchartrain. Had the barrier plan existed, it is plausible that its levees could have made things even worse by trapping surge waters that had either entered from Lake Borgne or had managed to overtop the barrier levees themselves.[33]

That was in 2005, when Katrina was still fresh in the public mind and a Republican Congress was eager to steer debate away from the Bush administration's bungled response. But more recently, the "Enviros Sank New Orleans" meme has been used to discredit the serious concerns environmental experts have regarding the Morganza-to-the-Gulf project, which I discussed in Chapter 5. For example, in a 2007 letter to the editor of the *Times-Picayune*, the leader of a levee advocacy group strongly denounced environmental opposition to the Morganza project. "Our group," she wrote," would like to give [Save Our Wetlands] a stern warning: Don't mess with our flood protection. We consider human life more precious than the environment. . . . [W]e will not allow a repeat of history from those who failed to learn from it."[34]

I agree that we should learn from the past. But what does the past teach? What message should we take from a previously forgotten lawsuit now more than 30 years old? In my view, there are two messages, both at odds with the skepticism expressed in that letter. First, natural infrastructure matters. At the time of the litigation, evidence suggested that the barrier plan might disrupt the estuary's salinity and natural tide flows, perhaps destroying the ecosystems of Lake Pontchartrain and three other connected water bodies. Those ecosystems were hardly trivial: they supported large commercial fisheries, hosted tourism and local recreation, and fed thousands of acres of storm-buffering swamps and marshes. As the reviewing court noted, even the chief engineer for the Corps's New Orleans Division had believed this and ordered updated studies (never completed) to make sure that recent changes to the plan's design would not cause the waters to stagnate.[35]

The barrier plan would also have walled off thousands of acres of protective wetlands so that they could be drained and populated with housing developments. As one example, the barrier option would have opened up 28,000 acres of marsh in eastern New Orleans, where developers planned to build a low-lying subdivision called Orlandia. Because the Corps later abandoned the barrier option, the land that would have become Orlandia instead became part of the Bayou Sauvage National Wildlife Refuge, a diverse array of freshwater marshes, estuaries, and hardwood hammocks that is billed as the largest urban wildlife park in the world. No picture of beauty today, much of Bayou Sauvage was demolished by Hurricane Katrina and will take years to regenerate. But that destruction reminds us of the force such ecosystems absorb and the protective capacity they offer. It also reminds us that sometimes landscape is best used as a place to visit, not live. Understanding natural infrastructure means paying attention to both the strengths *and* weaknesses of the local geography. Yet the barrier plan does not appear to have seriously considered either.

Why did the barrier plan ignore natural infrastructure? The answer ties back into conflicting goals and pork barrel politics. Faced with so many goals, including flood protection, promotion of commerce, environmental protection, and land development, the Corps relied on cost-benefit analysis to shape the priorities of the project. But the Corps's cost-benefit model valued the economic benefits of increased land value and new development over those of the preservation of natural systems or even the protection of human life. Part of the Corps's justification of its choice of the barrier option, one commentator noted, was that "[a]n extraordinary 79 percent [of the project's net benefits] were to come from new development that would be feasible with the added protection provided by the improved levee system."[36] This kind of development bias tends to gain support from local civic leaders who see such projects as a way to provide protection and pork at the same time. For these reasons, many community flood control projects are soon followed by new development in the flood-prone areas the projects have made available.[37]

Thus the second lesson of this chapter: that while it takes time and money to study local ecosystem effects, the work is not debilitating. In reality, NEPA did not stop the barrier project. Politics and money did. That project might have been justified if lives had been considered. To sum up: NEPA helps make *better* decisions in all long-term projects by forcing planners to consider environmental trade-offs and, just as important, forcing them to consider the ways protecting ecosystem functions might actually help realize a nonenvironmental goal.

10

PLANNING OUR FUTURES

When evaluating disaster risks, we need a new way to conceptualize the problem. We need a new map. Like the diagram of London's Underground, the map must be detailed enough to explain what matters most but concise enough to give practical guidance. The New Orleans Flood illustrates the danger of relying on the early twentieth-century versions of cost-benefit analysis, encapsulated in evaluations of Net Economic Development (NED). These analyses leave out nearly everything that really matters, including the protection of people, their homes, and their natural environment. But even if the Army Corps's analysis factored in monetized benefits of saved lives, avoided injuries, saved ecosystems, that reform would be insufficient. There are too many problems that cost-benefit analysis cannot correct for—social amplification, rational risk aversion, distributional unfairness, uncertainty, irreversibility, and discounting. Even serial defenders of cost-benefit analysis like Posner and Sunstein recognize many of these failings as they apply to low-probability, worst-case scenarios. Sunstein tries to shore up the cost-benefit project by mixing social narratives and safety margins with steel-spined utility, but one gets the sense that the utility side must always come first. That is the only way, really, to save the method from debilitating vagueness. But if you want to presume in favor of monetized numbers, you must put forward some evidence of their basic reliability in the context of disaster. Now we are in a Catch-22, for Sunstein has already told us that because of social amplification, irreversibility, and the like, these number are unreliable.

Feasibility

What about the precautionary approach? In areas of health, safety, and environmental protection, the feasibility principle has been relatively successful in pushing society toward more precaution while making sure the financial costs are not overly burdensome. Feasibility standards show great potential for disaster mitigation is certain areas, particularly where private development is concerned. Take building codes and construction standards. Many of these protections are not as aggressive as they should be. Earthquake standards, for instance, are commonly designed to avoid structural collapse but not to avoid damage beyond that point.[1] This is good news for the people inside the structures (who generally survive as a result) but not for property owners and insurers, who may pay millions or even billions of dollars in property damage that stricter codes would have prevented. (Even when a building does not collapse, it can be left in shambles.) A feasibility standard would reduce economic damage and the social costs of lost community by requiring codes demanding enough to avoid more damage but not so demanding as to debilitate the construction industry or distort the construction market. That is, a building code based on feasibility would maximize the reduction of risk within the constraints of cost-effective regulation. As technology improves over time and becomes cheaper, code standards can be ratcheted up, and the potential for damage can be reduced further. A feasibility approach could work for other standards, too, like house elevations in flood-prone areas or fire-resistance standards near dry forests.

But feasibility has its limits. For instance, this approach will not be effective in situations where basic safety requires more than what industry can provide in a cost-effective way without government help. And in situations where the government itself is providing the protection (by building levees or dams, for instance) a standard of cost-effectiveness based on private markets becomes highly problematic, if not irrelevant. Who is to say whether it is feasible for the federal government to invest $40 billion in flood control and coastal restoration? Because of the government's power to tax and to spend in deficit, such projects are always, in some sense, "affordable." But that does not mean the investment is wise.

Open-Ended Balancing

For large public works projects intended to mitigate disaster, we must rely on a different kind of precaution. In such cases, open-ended balancing seems appealing. In Chapter 8, we saw that Heinzerling and Ackerman recommend a version of this idea that they call "holistic" analysis. Posner recommends something like this for catastrophic situations where probability and severity are uncertain and nonquantifiable. Ackerman and Heinzerling's approach would entail consideration of much of the information included in a cost-benefit analysis. Scientific information on harms to humans and the environment and economic costs that are naturally stated in monetary terms would be incorporated into the holistic evaluation of the pros and cons of adopting one public policy over another. But the holistic approach would not take the further step of translating benefits, such as the saving of human lives, into dollars, and it would not be paralyzed in the absence of quantitative information about the probabilities of harm.[2]

Ackerman and Heinzerling would ratchet up the level of precaution where large uncertainty looms. In such cases they propose a version of the maximin, one inspired by economists Kenneth Arrow and Leonid Hurwicz, in which analysts would study the extremes of the range of possible outcomes and then compare the benefits and costs of responding to those extremes.[3] Ackerman and Heinzerling admit that sometimes the government will overspend, but they suggest that most of the time the costs would be worth it.

This balancing approach may prove too open-ended. Imagine cramming all of the data on past Gulf storms, sea-level rise, coastal subsidence, and greenhouse gases into a single report that bureaucrats, politicians, and lay members of the public could understand. One might do better with Borges's map! Even experts in the field risk being overwhelmed with data. Behavior studies show that as decision-makers confront tasks of greater and greater complexity, they tend to rely more heavily on cognitive biases, or shortcuts, to analyze information.[4] This suggests that unless facts are kept in some reasonable order, decision-makers might once again resort to false confidence in hard numbers or overemphasize the most salient examples.

A second problem involves the worst case itself. Recall the difficulty experts had in modeling the Standard Project Hurricane for use in designing the New Orleans levee projects (Chapter 9). Reliance on a single hypothetical storm or earthquake or volcanic explosion is another example of a shortcut we cannot afford where historic information is slim and uncertainties are high. In discussing nonlinear systems influenced by power laws, Farber explains the danger: "[In such circumstances] unlikely events . . . have a strong cumulative effect. If we focus only on what seem to be reasonably likely outcomes, we overlook the statistical possibility of nasty surprises. Worst-case analysis can be a useful reminder that through a string of unlikely coincidences, things may go very wrong indeed."[5] As a hedge against nasty surprises, Farber recommends looking instead at a "spectrum of possible 'worse' cases" to avoid focusing too heavily on only the most likely outcome.[6] Again, things appear to be getting too complicated. Is there a way that policy-makers can have the variety of values they need (economic and noneconomic), over a meaningful range of relevant possibilities, without getting lost in the weeds? It might be possible through the use of yet another decision-making technique: scenario planning.

Scenario Planning

Imagine, for a moment, that you are buying a car. You have several features you are looking for, not all of them compatible with one another. You would like a comfortable, spacious car, but you are also interested in fuel economy. Safety is an important concern, as is handling and appearance. You want reliability and a good price, too. To make your decision, you might create a list of these features, putting them in priority or even assigning each a certain weight. Some factors, like fuel economy, can be objectively translated into monetary terms; others, like safety or appearance, are more subjective. The priorities or weights you assign will help you evaluate trade-offs even if not everything is reducible to dollars and cents.

But things get more complicated. Let's imagine that there are a few possible, though unlikely, events that could occur in the next few years that would be affected by your choice of car. It's possible, for instance, that your boss might transfer you to the satellite office in the suburbs. That would make fuel economy a more valuable feature than it is today. Or your sister might have a baby. As a future doting uncle or aunt, you would want the car to be very safe and to have a comfortable backseat for transporting a little one. Let's assume you really don't have enough information to assign reliable probabilities to these events, so discounting the value of fuel economy or a comfortable backseat is not really an option—even if you knew how to use that Excel program. While it is true that any decision you make is not irreversible (you can always replace the car with a better option later on), the costs involved in correcting your choice are enough to make you think carefully about what you are doing. How do you take this uncertainty into account?

If you are like many people, you will develop, however informally, a set of future outcomes and match them with a possible car. There is, for instance, the scenario in which you buy the subcompact and next year your sister has twins (a bad combination). Or the one where you buy the Hummer, your job moves to the suburbs, and your sister relocates to Anchorage (worse). More moderate scenarios lie in between. Your objective here is not to get the *best* combination possible in *every* circumstance (you can't) but if possible to make a choice that is satisfying in most circumstances and tolerable under even the worst.

On a vastly different scale, this exercise is something like the challenge policy-makers face in designing hazard reduction initiatives. And the way many people would go about solving the problem, by imagining a few carefully chosen "possible futures," is a very simple example of what decision-making experts call scenario planning: the close examination of a carefully chosen range of potential futures. The method is popular in financial planning, military planning, and climate science, among other fields—all disciplines where predicting long-term activity is fraught with uncertainty. Unlike forecasts (think of your local weather report), scenarios do not try to predict what the future *will* look like but rather what it *could* look like.

Like the cost-benefit analysis of Ford's Whiz Kids, scenario planning grew out of the military bureaucracy, emigrated to the private sector, and then boomeranged back to government offices for civilian use. The method can be traced to the military's study of complex systems during World War II. During that time, interdisciplinary working groups made up of physicists, mathematicians, logicians, social scientists, and others developed techniques for navigating the storm of a fast-changing crisis. Using some of these insights, the RAND Corporation later pioneered a technique called "future-now" thinking, which aimed to analyze the needs for future weapons systems by combining rigorous analysis with imaginative thought experiments. Meanwhile, the Stanford Research Institute at Stanford University generated other important methods of long-range planning, incorporating ideas from business, economics, and military science.[7]

In the 1960s—with war in Vietnam and cultural unrest in America and Europe—global companies like General Electric and the Royal Dutch/Shell Group (Shell) became similarly interested in combining knowledge and imagination to peer more deeply into the future. Shell's impressive scenario planning exercises of the 1970s allowed it to anticipate the shift in economic power from oil companies to oil producers in the Middle East and the resulting instability in energy markets. Similar exercises enabled Pacific Gas and Electric, a utility in northern California, to respond effectively to the Loma Prieta earthquake and the system disruptions that followed. Even your blue jeans owe something to future-now thinking. Clothing manufacturer Levi-Strauss is a frequent user of scenario planning and has used it to study a variety of extreme scenarios, from the deregulation of American farms to the obliteration of the world's cotton crop.[8]

Government agencies around the world have also employed scenario planning to make high-stakes decisions in uncertain times. One of the best known examples comes from South Africa, when activists and governmental officials were negotiating an end to apartheid. In 1991—three years before the parties would reach agreement—a group of academics, politicians, administrators, trade unionists, and businesspeople met at Mont Fleur in South Africa to imagine what the country might look like in 2002. Facilitated by the head of scenario planning at Shell, participants considered dozens of political and economic scenarios, including revolution, government crackdowns, right-wing revolts, "free-market utopias," and more.[9] The group finally settled on four scenarios, chosen for their ability to illustrate the diversity and range of South Africa's plausible futures. The first, the so-called Ostrich scenario, described a country in which apartheid endured, the country remained an international pariah, and the economy collapsed. The Lame Duck scenario described a political transition that lacked speed and direction, giving birth to a weak economy and a dithering government. In the third scenario, Icarus, dramatic political and economic change would come quickly but would prove unsustainable and quickly fizzle. The last scenario, Flight of the Flamingos, described a course of democratic inclusion and balanced growth that might possibly keep the country on a sustainable path. It was this scenario that served as a model for political alignment during South Africa's constitutional negotiations and inspired the victorious African National Congress to moderate its economic policies.[10]

More recently, Finland's FINSKEN (short for "Finnish Scenario") project has developed integrated scenarios that can be used to analyze potential changes in environmental and economic factors that are important to the country's well-being. A crossdisciplinary effort coordinated by the Finish Environmental Institute, FINSKEN offers an interactive online database that allows users to map and graph scenarios in 30-year "time slices" from 2010 through 2099. The program allows users to consider plausible developments on many fronts, including economic development, acid rain, and sea-level rise.[11] Other countries that have used scenario planning to inform political and economic choices include Canada, Guatemala, Norway, Scotland, and the United States.[12] Municipalities and other local governments around the world now also take advantage of these techniques.

KNOWLEDGE AND IMAGINATION

Albert Einstein famously believed that "imagination is more important than knowledge."[13] Without taking sides, scenario planning endeavors to improve both. It improves knowledge in three ways. First, scenario planning improves knowledge by avoiding the urge to predict or forecast. Where uncertainty is high and time-lines are long, forecasting often means failure. It is not just that in such circumstances, any single prediction is bound to be wrong, but that in the whirl of feedback loops and nonlinear progressions, there is a good chance it will be *very* wrong. Cost-benefit approaches provide poor measures when they depend on the forecasting of too many long-term and uncertain events. Monetizing the value of one preserved coastal swamp—if done with any desire for comprehensiveness—would require speculations about the future real estate market; future population patterns; future sea-level rise; future storm trends; future seafood yields; future seafood markets; future recreational, aesthetic, and spiritual trends; and much more. There is little evidence that experts can come anywhere near accurate predictions about many of these things 50 or 100 years out.

Second, scenario planning broadens knowledge by taking a holistic approach to describing the circumstances. Scenario plans often include information about economic and demographic trends, weather and climate changes, and shifts in technology. This information is described with reference to narrative descriptions as well as numbers. The strong emphasis on narrative allows the technique to capture a problem in its full complexity. In contrast to cost-benefit analysis, scenario plans do not monetize eventualities but allow them to be studied on their own terms. Narratives are also more easily understood by laypeople and experts without economics training. In addition, narratives tend to be more memorable (or as cognitive psychologists say, "available"), making it harder for planners to dismiss them simply because the events they describe are less likely than other scenarios.[14] Narrative scenarios can also be designed to change as circumstances on the ground change, offering alternative snapshots of the future at any given point in time. The dynamic and adaptive nature of scenario planning allows decision-makers to adjust laws and public policy to accommodate new possibilities.

Third, by emphasizing system structure, scenario planning forces planners to consider endogenous as well as exogenous forces.[15] Other methods of risk management too often allow planners to neglect the possibility of change within society itself. Cost-benefit analysis, for instance, has no uniform way of accounting for the possibility that technological improvements in the future might reduce regulatory costs (by making compliance easier) or, conversely, reduce regulatory benefits (by making people less vulnerable to the regulated harm). Too often, technological improvement is not seriously considered at all. Feasibility analysis suffers the same problem where speculations about compliance costs are concerned. Traditional disaster projections, like the Standard Project Hurricane, similarly fail to consider such important societal changes as increased population, coastal development, or escalating poverty. In contrast, scenario planning requires decision-makers to consider the external forces of storms, quakes, and slides in conjunction with important economic and societal patterns. By spotlighting issues of public exposure and social vulnerability, the technique is particularly useful for advocates of disaster justice.

Finally, by pulling in experts from many fields and describing events in terms that all can understand, scenario planning creates a system of redundancies or "double checks" in which contributors are encouraged to review and rethink other people's conclusions. Such redundancy leads to a more complete and consistent result.[16]

Equally important, scenario planning forces decision-makers to use their imaginations. The very process of constructing scenarios stimulates creativity among planners, helping them to break out of established assumptions and patterns of thinking. Scenario planning is also more proactive than cost-benefit analysis. That is, rather than evaluating the situation as statistics say it is likely to be, scenario planners are encouraged to conceive of dangers and failures that may not have even been identified yet. Experts in organizational structure say that failures in crisis management can usually be attributed to a lack of imaginative vision or denial of that vision.[17]

In September of 2001, when most antiterror experts were focused on the "suitcase bomb," America and its leaders were stunned to see hijacked planes used as weapons of mass destruction, a possibility that most people in the intelligence communities had seldom considered.[18] The 9/11 Commission Report later found that although intelligence officials were on notice throughout the summer of 2001 that terrorist activity was high, they were never able to assemble the evidence they were collecting into a recognizable plan. As one CIA operative testified, "no one looked at the bigger picture; no analytic work foresaw the lightning that could connect the thundercloud to the ground."[19] They weren't the only ones unprepared. The engineering and emergency response communities in New York had never envisioned the collapse of the World Trade Center, and consequently had never prepared for it. Similarly, emergency planners had never imagined an event so large as to completely overwhelm the city's capacity to respond.[20]

Failures of vision also occurred in the years before Hurricane Katrina. Recall that experts designing the Standard Project Hurricane dropped Hurricane Camille and the Keys Storm from their historical models because they were not considered "reasonably characteristic" of the region (Chapter 9). Even when planners do envision the right scenario, an overemphasis on probability can force them to deny what their minds see. This happened in the events leading up to Katrina, too. In July 2004, federal, state, and local planners considered an imagined worst-case scenario for New Orleans called Hurricane PAM. The event was remarkably similar to Hurricane Katrina, envisioning overtopped levees, a flooded city, and tens of thousands of people in need of emergency relief. But decision-makers dismissed the scenario as too extreme to be given any priority. Apparently, the exercise was neither developed nor implemented in a way to move planners beyond the threshold of probability.[21]

Hurricane PAM shows that scenario planning is not without pitfalls. The method relies on the subjective decisions of planners who choose the range of futures and the ways they are described. In the best applications, scenarios are chosen to be complementary and in some ways overlapping. If the differences among scenarios are too stark, one or more may be viewed by the decision-makers as being too far outside the norm and dismissed out of hand (a version of the "bias toward zero"). There is also a tendency for people, particularly nonexperts, to bet on a particular scenario rather than to see the offerings as a range of possibilities for which one should be prepared. Similarly, nonexperts will sometimes lock into a menu of possibilities generated at one point in time and resist changes in course, even when the menus have changed to accommodate new information or more recent developments. Conducted properly, scenario planning is also resource intensive, requiring the cooperation of experts in many fields and often the participation of business leaders, community groups, and other members of the public.

At this point, scenario planning, as a discipline of study remains vastly underresearched. Many assertions about scenario planning remain at the theoretical stage and have not been tested by rigorous empirical analysis. In addition, much of the theorizing and anecdotal evidence comes from the private sector. Experts know much less about how scenario planning should be successfully integrated into government settings.[22] Yet for many kinds of disaster mitigation, scenario planning could prove extremely useful. For this reason, governments and universities should ramp up research in this area. For now, we can gain some insight by looking at a few applications of scenario planning in the environmental and disaster contexts.

ENVIRONMENTAL SCENARIOS

The scenario approach is not new to environmental law and policy. The federal EIS process set forth in the NEPA, is in effect a scenario planning device. When a federal agency plans major action that could significantly affect the environment, the EIS process requires the agency to describe the affected environment, present a reasonable range of alternatives (including no action), and to analyze the impacts of each alternative. Significantly, NEPA does not allow agencies to ignore potential threats simply because the risks cannot be quantified. Regulations implementing the statute instruct that where important information is not available at a reasonable cost, an agency must include in its EIS, among other things, "a summary of existing credible scientific evidence which is relevant to evaluating the *reasonably foreseeable* significant adverse impacts on the human environment; and . . . the agency's evaluation of such impacts based upon theoretical approaches or research methods generally accepted in the scientific community."[23] In this case, "reasonably foreseeable" includes impacts that "have catastrophic consequences, even if their probability of occurrence is low, provided that the analysis of the impacts is supported by credible scientific evidence, is not based on pure conjecture, and is within the rule of reason."[24]

More could be done to make NEPA an even better planning tool where catastrophe is concerned. Farber has argued that agencies should be told to be much more explicit about uncertainties lurking within their models. Where catastrophic outcomes are credible, agencies should be required to consider reasonable margins of safety or other precautions, regardless of the likelihood of occurrence. If the future outcome is important, agencies should continue to monitor developments even after the predictions are made. Such ongoing study would allow government to update its assumptions when facts change and would also provide a means for testing agencies' predictive capacities.[25]

More sophisticated versions of scenario planning are used in various U.N. projects to assess global environmental risks. The U.N. Environmental Programme's Millennium Ecosystem Assessment, which aims to assess the health of the world's natural infrastructure, uses an array of scenario planning techniques, some newly developed, others borrowed from previous methods pioneered by the U.N. Intergovernmental Panel on Climate Change, the World Water Council, and the well-respected Global Scenario Group. To show the level of ambition, the Millennium Ecosystem Assessment aims to

> develop scenarios that connect possible changes in drivers (which may be unpredictable or uncontrollable) with human demands for ecosystem services. The scenarios will link these demands, in turn, to the futures of the services themselves and the aspects of human welfare that depend on them. The scenario building exercise will break new ground in several areas: development of scenarios for global futures linked explicitly to ecosystem services and the human consequences of ecosystem change, consideration of trade-offs among individual ecosystem services within the "bundle" of benefits that any particular ecosystem potentially provides to society, assessment of modeling capabilities for linking socioeconomic drivers and ecosystem services, and consideration of ambiguous futures as well as quantifiable uncertainties.[26]

Echoing a point made in the NEPA regulations, the authors write: "The credibility of assessments is closely linked to how they address what is not known in addition to what is known. . . . As part of any assessment process, it is crucial to estimate the uncertainty of findings even if a detailed quantitative appraisal of uncertainty is unavailable."[27]

DISASTER SCENARIOS

We can see the beginnings of such an approach in the way the U.S. Army Corps of Engineers is addressing storm protection in post-Katrina New Orleans. Spurred by local outrage and some embarrassment, Congress directed the Corps to examine the possibility of upgrading Louisiana's storm protection system (which would include both levee improvements and the restoration of protective coastal wetlands) to defend against some level of

Category 5 hurricane (the strongest category on the Saffir-Simpson scale).[28] Further, Congress instructed that the Corps is not to be bound by the NED method, its traditional cost-benefit approach. In response, the Corps has developed a new "risk-informed" framework that abandons some of the familiar trappings of cost-benefit analysis in favor of a more holistic approach.[29] The framework requires examining a range of storm scenarios from weak to strong, with as much quantitative data as possible about storm dynamics, engineering systems, land use, and population centers. It will then seek to predict the effects under each scenario, from loss of life to loss of infrastructure. Distributional effects can be pinpointed by neighborhood using geographic information system technology. Significantly, this framework does not envision the monetization of nonmonetary losses such as loss of lives or degraded ecosystems, nor does it discount such future losses into the present. Rather, all the elements are put on the table on their own terms: dollars, lives, acres of wetlands, and so on. The various scenarios can be matched with several varieties of storm protection in order for decision-makers to determine what system is most desirable. As currently conceived, the choice of protection would be made by Congress, a democratically elected body. These developments are good news for cost-benefit skeptics; but they will put the democratic process to the test. Questions abound. How will this mass of information be organized and communicated to legislators and the general public?

Some answers are suggested by the draft version of the Army Corps's "Louisiana Coastal Protection and Restoration Technical Report" that as of this writing is under review with the National Academy of Sciences.[30] The report combines two frameworks for organizing information that are not unlike those used in the hypothetical situation of buying a car. The first framework, "multiple-criteria decision analysis," is used to identify and weigh the criteria, or "metrics," that we most care about in a storm-protection system. Some metrics can be expressed easily in numbers; others cannot. In the draft report, these metrics include things like life-cycle cost, construction time, positive and negative impacts on wetlands, protection of people, and protection of historic districts.[31] Weights are assigned to the criteria through an involved process of stakeholder meetings that involve dozens of groups, including community associations, environmental organizations, and oil companies.

Once priorities are understood, they are considered against a second framework based on scenario planning. Scenarios will describe a range of possible futures based on potential levels of relative sea-level rise 50 years from now and potential levels of regional redevelopment. One scenario might show a future where the sea level has risen only moderately and the regional population has not significantly grown (a good scenario from the point of few of safety). Another would show a future of high sea-level rise and sprawling growth along the coast (a bad scenario).[32]

Then planners will posit for each of those futures a given level of potential storm, from a 10-year storm (with a 10 percent chance of occurring in a given year) all the way to a 2,000-year storm (with a 0.05 percent chance of occurring in a given year). In each of these scenarios, planners will show the predicted losses in terms of residential and nonresidential structures, contents of structures, vehicles, farmland, transportation costs, emergency response costs, and so on. The analyses will go through repeated public comments and hearings as well as multiple peer review. The hope is to make the process as transparent as possible and to be up front about the level of uncertainty as well as the value choices that are inherent in the process.

It is too early to know how effective this method will be. The process obviously requires a careful balance between giving decision-makers enough information to make meaningful choices without overwhelming them with so many facts and figures that cognitive shortcuts take over. Obviously, where the public is involved, scenarios must be kept at some easily manageable level. That is one reason, for instance, that the possible futures are confined to only two levels of possible sea-level rise and development. Behavioral psychologists have just begun uncovering the mysteries of how the brain receives and analyzes large packages of information. Continued research in this area is the key to improving the effectiveness of scenario planning. The process is also extremely involved and resource intensive. An agency should not be expected to employ such an approach where the stakes are low and the level of confidence in an outcome relatively high.

Innovations like multiple-criteria decision analysis and scenario planning should prompt policy-makers to ask if such models should be used for other regions of the country where disaster is a major concern. Many cities, for instance are considering public works projects to defend themselves from climate-induced sea-level rise. If risk-informed decision-making makes sense for New Orleans flood protection, does it also make sense in San Diego and Miami? Should it be applied to the 36 weakened levees the Army Corps has identified in California's Sacramento District?[33] In addition, policy-makers must consider the best way to protect risk-informed decision-making from the centralized cost-benefit review at OMB. If the Army Corps determines under risk-informed analysis that a levee should be designed in a specific way to protect the surrounding population, it should not fear interference from a White House agency that is using a different measuring stick. Who says the debate over cost-benefit analysis has already been won? Where disasters are concerned, the conversation in America is just beginning.

CONCLUSION

On a crisp November evening in 2007, the Lower Ninth Ward came back to life, if only temporarily. Hundreds of cars sat parked along vacant lots among twisted signs and broken concrete. People from all parts of the city were outside milling about. Caterers handed out free gumbo in a large white tent, and the Rebirth Brass Band blasted out a euphoric set of New Orleans standards. Gradually folks eased their way toward several rows of bleachers that had been situated to face the intersection of two rutted streets. A forest of weeds waved in the background. The occasion was an evening of outdoor theater arranged by the Classical Theater of Harlem and a team of local actors. The play everyone had come to see was *Waiting for Godot*. In art, perfect moments do not come along often. But this was one. A city that in the new millennium had become a metaphor for agonizing delay was now seeking solace in a vaudeville-inflected play that 50 years ago had posited waiting as the core human experience.[1]

There is waiting before and after a disaster. We know the latter best, having watched television accounts of the rescue and recovery efforts following tragedies like the Asian Tsunami, Hurricane Katrina, and Cyclone Nargis. The waiting *before* a disaster is less dramatic and harder to describe. Think of it as that vague sense of anxiety people sometimes feel—like a virus running in the background of a computer program—on learning that their community is living on borrowed time. Many New Orleanians felt that way in the years before Katrina. In cities like San Francisco, Sacramento, Houston, and Miami, some people feel that way now.

The remedy for this condition is action. That's what this book is about. I have suggested that those seeking to improve disaster policy might glean some lessons from environmental law and that those lessons can be expressed as three principles: Go Green, Be Fair, and Keep Safe. Protecting people from natural hazards requires a steady commitment to protecting and maintaining the natural systems that keep us secure. In addition, we must make special efforts to reduce the levels of exposure and vulnerability that many less advantaged groups face in our country and in the rest of the world. Finally, our laws must take a precautionary and holistic approach to risk management.

In making my case, I have tried to build carefully on theoretical foundations. Each of these three lessons follows from a belief that law must serve more than free-market principles. To be sure, markets produce an array of benefits. But we make a grave error when we equate the necessities of the market with the common good. In such circumstances, protective wetlands and hillside forests are exploited beyond repair. Emergency response and recovery degenerates into do-it-yourself survival. And protecting society from low-probability, high-stakes events becomes a game of averages and hypothetical auctions. In contrast, I urge that legal reforms in disaster policy be marked by compassion and a philosophical belief that all people and communities are entitled to some basic level of attention and safety, simply by virtue of their membership in society.

But disaster law, like environmental law, is an applied discipline. With that in mind, I have tried to offer some ideas for concrete policy change. On the subject of natural infrastructure, for instance, I've recommend a scientific inventory of forest-based ecosystem services, a presumption that favors nonstructural methods of flood control, a "despoilers pay" attitude toward infrastructure restoration, and a relationship approach to property law that would make it easier for plaintiffs to use the court systems to prevent damage to protective natural features on private land.

In the pursuit of greater fairness—"disaster justice," as I call it—I recommend an array of domestic and international reforms. We should shore up our traditional laws protecting health, welfare, and the environment, including laws that protect communities from toxic releases, an occurrence that is too common during extreme events. The Stafford Act should be amended to include private remedies for discriminatory acts and a right of action to urge state and local compliance with planning requirements. I also call for an executive order on disaster justice that would seek to integrate disaster justice values into the planning of all federal agencies, as well as legislation to govern the use of large-scale compensation funds. On the international side, I argue that a set of binding principles should be negotiated to provide some minimum protection for vulnerable populations in times of catastrophe, based on the Hyogo Framework and the Guiding Principles on Internally Displaced Persons.

In the interest of public safety, as well as fairness, I would replace the cost-benefit models used in disaster mitigation with precautionary models. I have suggested feasibility standards for building codes and development projects designed to protect people and communities as much as is technologically and economically possible. For large public works projects, I've recommended a precautionary approach based on open-ended balancing. In contexts of nonquantifiable uncertainty where a disaster's probability is low but its damaging effects severe, I have suggested that agencies pursue more innovative models of scenario planning, inspired by the U.N. Millennium Ecosystem Assessment or the Army Corps's scenario planning approach to storm protection on the Gulf Coast. This is by no means an exhaustive list of what we might try. Surely better and even grander ideas are on the horizon.

In addition to all this, we cannot forget climate change, that thundering elephant in the room. Not only might global warming increase the frequency or severity of some events, like tropical storms, it promises to make their occurrence and effects less certain. The vulnerability of societies to such damage is not uniform. Countries in tropical areas are more likely to be hit harder. Tropical nations already endure climatic disturbances such as cyclones, monsoons, and the El Niño and La Niña cycles. There is persuasive evidence that global warming could extend or intensify these events.[2]

As the 2004 Asian Tsunami demonstrated, people in the developing world are least able to cope with sudden disasters when they do occur because of substandard infrastructure, poor emergency response, and insufficient medical services. Indeed, research suggests that the death tolls of a natural disaster in a given nation are generally linked to its economic welfare and degree of income inequality.[3] Developing countries will also have greater sensitivity to climate change because of their economic reliance on agriculture and fragile ecosystems. In Sub-Saharan Africa, for instance, 64 percent of the people are employed in the rural farming sector.[4] The percentage is nearly the same in South Asia.[5] People in developing countries also rely more heavily on subsistence goods and services drawn from the environment, such as fish, vegetables, firewood, fiber, and drinking water.

In developing countries, climate victims will be mainly people of color. A large share of them will be infants and children, elderly people, and the disabled.[6] This means more work for women and girls, who generally care for the children, the elderly, and the infirm.[7] Women and girls, who in many cultures also grow the food and collect the water, will also find these chores consuming more of their time, leaving less time for education and vocational training.[8] By pushing girls out of school, climate change is expected to increase female illiteracy.[9]

Wealthy countries like the United States will also face serious climate-related problems, including sea-level rise and more intense droughts and storms. In the United States, state governments and private research centers have begun documenting an array of climate-induced harms expected in California, the Pacific Northwest, and the Northeast. A 2008 government report summarizing the effects of climate change in the United States warns of devastating water shortages in the Southwest, thousands of miles of washed-out highways and sunken railroad lines in the East, and an alarming increase in heatstroke, smog-induced asthma, and posttraumatic stress disorder.[10] If protective action is not taken in the next 50 years, hundreds of miles of storm-buffering wetlands in the Gulf and along the Atlantic coast could be swallowed up by sea-level rise, along with recreational beaches, coastal spits, and barrier islands. Hurricane Katrina reminds us that even in rich countries, weather-related violence visits particular harm on vulnerable classes—the poor, the infirm, the elderly, and children. The government's own reports on climate change emphasize this fact, warning that vulnerable classes will likely bear the brunt of increased heat waves, flooding, and drought.[11]

In the face of such challenges, it is tempting to throw up your hands and wonder what to do; or to convince yourself that the right answer will come if only you wait patiently enough. But waiting in cases like this leads to further disappointment. That's why for most people *Waiting for Godot* is such a downer. We remember the pointless milling around, the repeated "Yes, let's go," with nothing happening afterward; it's a story of apathy, despair, and sore feet. But deep inside that void is energy—and courage. I think back to that evening in the Lower Nine when Vladimir—an exhausted, middle-aged black man, dressed in a rumpled jacket and a dust-covered bowler—stood against the apocalyptic tide and roared:

> Let us not waste our time in idle discourse! Let's do something, while we have the chance! It's not every day that we are needed. . . . [A]t this place, at this moment of time, all mankind is us, whether we like it or not. Let us make the most of it before it is too late![12]

NOTES

INDEX

NOTES

Introduction

1. My discussion of the Lisbon Earthquake draws from Nicholas Shrady, *The Last Day: Wrath, Ruin, and Reason in the Great Lisbon Earthquake of 1755* (New York: Viking, 2008) and T. D. Kendrick, *The Lisbon Earthquake* (Philadelphia: Lippincott, 1957).

2. Oliver Wendell Holmes, "First of November: The Earthquake Day," (undated) (second ellipsis in the original), www.phenomenon.org (accessed Jan. 16, 2009). One of Holmes's best known poems, "The Deacon's Masterpiece," offers a less serious take on that earthquake. See Oliver Wendell Holmes, "The Deacon's Masterpiece, or The Wonderful One-Hoss Shay" in *Illustrated Poems of Oliver Wendell Holmes* (1884) (Ann Arbor. University of Michigan Library, 2006), 68.

3. See, generally, John Barry, *Rising Tide: The Great Mississippi Flood of 1927 and How It Changed America* (New York: Simon and Schuster, 1998).

4. See, generally, John McQuaid and Mark Schleifstein, *Path of Destruction: The Devastation of New Orleans and the Coming Age of Superstorms* (New York: Little, Brown, 2006).

5. R. B. Seed et al., "Investigation of the Performance of the New Orleans Flood Protection Systems in Hurricane Katrina on August 29, 2005" (draft final report), June 1, 2006, vii (attributing the flood, in part, to "the poor performance of the flood protection system, due to localized engineering failures, questionable judgments, errors, etc. involved in the detailed design, construction, operation and maintenance of the system"), www.ce.berkeley.edu (accessed Jan. 12, 2009); Ivar Ll. van Heerden et al., "The Failure of the New Orleans Levee System during Hurricane Katrina," Dec. 18, 2006, (noting failures in worst-case storm analysis, levee design and construction, and levee maintenance), iii–ix, www.publichealth.hurricane .lsu.edu (accessed Jan. 12, 2009); U.S. Army Corps of Engineers, "Performance Evaluation of the New Orleans and Southeast Louisiana Hurricane Protection System: Draft Final Report of the Interagency Performance Evaluation Task Force" (June 2006), 3 (citing poor design and construction as the reason for levee failures) ipet.wes.army.mil (accessed Jan. 12, 2009).

6. See, e.g., G. van der Vink et al., "Why the United States Is Becoming More Vulnerable to Natural Disasters," *EOS, Transactions, American Geophysical Union*, 79 (1998): 533–537.
7. See James Nachtwey, "Henri Cartier-Bresson, 1908–2004," *Time*, Aug. 8, 2004 (quoting Cartier-Bresson), www.time.com (accessed Jan. 10, 2009).
8. U.N. International Strategy for Disaster Reduction, "Terminology on Disaster Risk Reduction," www.adrc.or.jp (accessed Jan. 16, 2009).
9. 42 U.S.C. sec. 5122.
10. See, generally, Daniel A. Farber et al., *Disaster Law and Policy* (New York: Aspen, 2009).

1. Natural Infrastructure

1. The park contains 20,020 acres. "Jean Lafitte National Historical Park and Preserve," www.nps.gov (accessed Jan. 12, 2008).
2. For more on the history of this area, see generally Frederick Turner, "Barataria," *American Heritage*, August/September 1980, 35; D. R. Levin, "Transgressions and Regressions in the Barataria Bight Region of Coastal Louisiana." *Gulf Coast Association of Geological Societies Transactions*, 41 (1991): 408–431.
3. Bradford Torrey and Francis H. Allen, eds., *The Journal of Henry David Thoreau: In Fourteen Volumes Bound as Two Volumes, I–VII (1837–October, 1855)* (New York: Dover, 1962), 102. For sublime analysis of Thoreau's work as it relates to modern ecology, see Daniel B. Botkin, *No Man's Garden: Thoreau and a New Vision for Civilization and Nature* (Washington, D.C.: Island Press, 2001).
4. Henry David Thoreau, "Walking," in *Excursions* (Cambridge, Mass.: Riverside Press, 1863), 280.
5. Ibid.
6. Richard Louv, *Last Child in the Woods: Saving Our Children from Nature-deficit Disorder* (Chapel Hill, N.C.: Algonquin Books, 2006).
7. See, e.g., C. C. Lockwood, *Atchafalaya* (Baton Rouge: Louisiana State University Press, 2007), and *Still Waters: Images 1971–1999* (Baton Rouge: Louisiana State University Press, 2000); Jack and Anne Rudloe, "Louisiana's Atchafalya; Trouble in Bayou County" (photographs by C. C. Lockwood), *National Geographic*, September 1979, 377–397.
8. C. C.'s journals and Sue's lesson plans are available at Marsh Mission, www.marshmission.com (accessed June 6, 2008).
9. C. C. Lockwood and Rhea Gary, "Houseboat Journey through the Wetlands," in *Marsh Mission: Capturing the Vanishing Wetlands* (Baton Rouge: Louisiana State University Press, 2005), 1.
10. Percy Viosca, Jr., "Louisiana Wet Lands and the Value of Their Wild Life and Fishery Resources," *Ecology* 9 (1928): 216–229, 221. My discussion of Viosca's methodology and conclusions come from this article.

11. My calculation of relative value is based on the Consumer Price Index for 2008.

12. Viosca, "Louisiana Wet Lands," 221.

13. Ibid., 227.

14. Ibid., 228.

15. Twenty-five percent of the nation's coastal wetlands reside in southern Louisiana. Mike Tidwell, *Bayou Blues: The Rich Life and Tragic Death of Louisiana's Cajun Coast* (New York: Pantheon, 2003), 6.

16. Eric Berger, "Keeping Its Head above Water: New Orleans Faces Doomsday Scenario," *Houston Chronicle*, Dec. 1, 2001. But see Kristina Hill and Jonathan Barnett, "Design for Rising Sea Levels," *Harvard Design Magazine* (Fall 2007/ Winter 2008): 5 (stating that every 5 miles of coastal wetlands could reduce storm surge by about a foot).

17. Ibid.

18. Ibid.

19. Ibid.

20. Oliver A. Houck, "Land Loss in Coastal Louisiana: Causes, Consequences, and Remedies," *Tulane Law Review* 58 (1983): 3–167, 84–86; Mike Tidwell, *Bayou Farewell: The Rich and Tragic Death of Louisiana's Cajun Coast* (New York: Vintage, 2004), 134.

21. Edward D. Houde and Edward S. Rutherford, "Recent Trends in Estuarine Fisheries: Prediction of Fish Production and Yield," *Estuaries* 16 (1993): 161–176, 169.

22. About 70 percent of all birds that migrate through the United States use the Mississippi and Central flyways. U.S. Army Corps of Engineers, *Louisiana: Ecosystem Restoration Study, Final, Louisiana Coastal Area* 1 (2004): B 1.1, www.lca.gov (accessed July 1, 2008). The coastal plain also supports several endangered and previously endangered species, including bald eagles, brown pelicans, alligators, and various kinds of whales. Houck, "Land Loss," 90. The birdlife moving through southern Louisiana supports significant commercial enterprises, including tourism, birding, and hunting; 88–90.

23. Houck, "Land Loss," 78–79. The marshes' natural store of freshwater also acts as a bulwark against intruding saltwater, which, were it allowed to flow uninhibited up the bayous, would destroy crucial shellfish habitat and poison groundwater supplies south of New Orleans; 80–81.

24. Ibid., 99 (estimating an annual value of around $10 billion in 1983, using two different valuation methods).

25. U.S. Army Corps of Engineers, *Louisiana: Ecosystem Restoration Study*, B 1.1, iii. In the 1970s, Louisiana was losing an estimated 25,200 acres per year from a combination of natural and human process; from 1990 to 2000, the rate slowed to 15,300 acres per year; ibid.

26. That loss would represent 10 percent of Louisiana's remaining coastal plain. Ibid., B 1.1, iii.

27. Ibid., B 2.1.1.4.

28. Houck, "Land Loss," 15.

29. Hydraulic forces erode the banks of such canals, causing them to widen at sometimes alarming rates. The surface area of the coast's artificial waterways may itself account for "two to four percent of [the coast's] total land mass." Ibid., 37.

30. Ibid., 39–40.

31. Tidwell, *Bayou Farewell*, 131–132.

32. Ibid.

33. Louisiana Coastal Wetlands Conservation and Restoration Task Force and the Wetlands Conservation and Restoration Authority, Louisiana Department of Natural Resources, *Coast 2050: Toward a Sustainable Coastal Louisiana* (1998), www.lcoast.gov (accessed July 2, 2008).

34. Tidwell, *Bayou Farewell*, 134.

35. Houck, "Land Loss," 88–90; Michael Scherer, "Bush Fought Funding in Energy Bill for Gulf Coast Protection," *Salon Online*, Sept. 1, 2005, www.salon.com (accessed June 23, 2008).

36. Mark Schleifstein, "Land Lost," *New Orleans Times-Picayune*, Oct. 11, 2006.

37. Ibid.

38. Ian McNulty, "Pearl Jammed," *Gambit Weekly*, Dec. 9, 2008, 32.

39. "$38 Million Deal Reached in Minnesota Bridge Collapse," May 2, 2008 www.cnn.com (accessed Jan. 19, 2010).

40. U.S. Army Corps of Engineers, *Levees of Maintenance Concern*, (Feb. 1, 2007), www.hq.usace.army.mil (accessed June 20, 2008).

41. See ibid. (showing 48 levees with an "unacceptable" risk as being in California, mostly in the "Sacramento District").

42. Adam Cohen, "Public Works: When Big Government Plays Its Role," *New York Times*, Nov. 13, 2007, A28.

43. Nick Miroff, "Collapse Spotlights Weaknesses in U.S. Infrastructure," *Washington Post*, August 3, 2007, A08.

44. *Webster's New World Dictionary*, ed. Michael E. Agnes, (New York: Wiley, 2003).

45. U.S. Government Accountability Office, *Critical Infrastructure Protection: Progress Coordinating Government and Private Sector Efforts Varies by Sectors' Characteristics*, GAO-07-39 (Washington, D.C.: October 16, 2006): 17.

46. Kanchan Chopra et al., eds., *Ecosystems and Human Well-Being: Policy Responses*, vol. 3 (Washington D.C.: Island Press, 2006), 27.

47. I am not the first to use the term *green infrastructure* in this sense. See Mark A. Benedict and Edward T. McMahon, Green Infrastructure: Smart Conservation for the 21st Century, 5, Washington, D.C.: Sprawl Watch Clearinghouse, 2001, available at www.sprawlwatch.org (accessed Jan. 19, 2010). ("[G]reen infrastructure is the ecological framework needed for environmental, social and economic sustainability—in short it is our nation's *life sustaining system*.")

48. Casey Jarman and Robert R. M. Verchick, "Beyond the 'Courts of the Conqueror': Balancing Private and Cultural Property Rights under Hawaiian Law," *Scholar: St. Mary's Law Review on Minority Issues* 5 (2003): 204–205.

49. *Gibbons v. Ogden*, 22 U.S. 1 (1824).

50. See, e.g., Edward Farnworth et al, "The Value of Natural Ecosystems: An Economic and Ecological Framework," *Environmental Conservation* 8 (1981): 275–282; Robert Costanza, ed., *Ecological Economics: The Science and Management of Sustainability* (New York: Columbia University Press, 1991); Gretchen C. Daily, ed., *Nature's Services: Societal Dependence on Natural Ecosystems* (Washington, D.C.: Island Press, 1997). For more on the development of ecological economics as a discipline, see J. B. Ruhl, Steven E. Kraft, and Christopher L. Lant, *Law and Policy of Ecosystem Services* (Washington D.C.: Island Press, 2007).

51. Spencer Weber Waller and Brett M. Fischman, "Essential Facilities, Infrastructure, and Open Access," working paper, Loyola University of Chicago School of Law, Chicago, Illinois, 2006).

52. Ibid.

53. Christine Harvey, "Shaw Group Throws in Towel on Causeway Deal," *New Orleans Times Picayune* Mar. 26, 2008.

2. Our Incredible Shrinking Infrastructure

1. Heather Tallis and Peter Kareiva, "Ecosystem Services," 15, no. 17 (2005): R746–R748, R747.

2. Millennium Ecosystem Assessment, *Ecosystems and Human Well-Being: Synthesis* (Washington D.C.: Island Press, 2005), 20.

3. Ibid., 6.

4. Ibid., 41–45, table 2.1.

5. Janet Abramowitz, *Unnatural Disasters*, ed. Linda Starke (Washington, D.C.: World Watch Institute, 2001), 7.

6. Karen Sudmeier-Rieux, Hillary Masundire, Ali Rizvi, and Simon Rietbergen, eds., *Ecosystems, Livelihoods and Disasters: An Integrated Approach to Disaster Risk Management* (Gland, Switzerland: IUCN Press, 2006), ix.

7. Abramowitz, *Unnatural Disasters*, 41.

8. Ibid.

9. Sudmeier-Rieux et al., *Ecosystems, Livelihoods and Disasters*, 7.

10. U.S. Army Corps of Engineers, *Performance Evaluation of the New Orleans and Southeast Louisiana Hurricane Protection System: Draft Final Report of the Interagency Performance Task Force* (June 2006), ipet.wes.army.mil (accessed July 2, 2008).

11. Jeff Hecht, "Disappearing Deltas Could Spell Disaster," *New Scientist*, Feb. 20, 2006, 8.

12. Ibid., p. 8; Graham Bowley, "Cyclone Toll Exceeds 3,100 in Bangladesh," *New York Times,* Nov. 20, 2007, A10.

13. Donald F. Boesch and Nancy N. Rabalais, eds., *Long Term Environmental Effects of Offshore Oil and Gas Development* (London: Taylor Francis, 1987).

14. U.S. General Accounting Office, *Alaska Native Villages: Villages Affected by Flooding and Erosion Have Difficulty Qualifying for Federal Assistance* (June 20, 2004), www.gao.gov (accessed July 2, 2008).

15. Ibid., pp. 1–10; Emma Schwartz, "Attacking Climate Change in Court," *U.S. News and World Report,* Mar. 13, 2008, 12; Complaint for Damages, Native Village of Kivalina v. ExxonMobil, et al., CV 08–1138 (N.D. Calif., Filed Feb. 26, 2008).

16. Louisiana Sea Grant, "Louisiana Hurricane Resources," www.laseagrant.org (accessed July 2, 2008).

17. R. S. De Groot, M. A. M. Stuip, C. M. Finlayson, and Davidson, *Valuing Wetlands: Guidance for Valuing the Benefits Derived from Wetland Ecosystem Services,* Ramsar technical report no. 3/CBD technical series no. 27. Montreal: Ramsar Convention Secretariat, (2006), 28, 27 (fig. 7), 26 (table 11).

18. Daniel M. Alongi, "Present State and Future of the World's Mangrove Forests," *Environmental Conservation* 29 (2002): 331–349, 331–332.

19. Sue Wells et al., *In the Front Line: Shoreline Protection and Other Ecosystem Services from Mangroves and Coral Reefs* (UNEP, 2006), 15; Sudmeier-Rieux et al., *Ecosystems, Livelihoods and Disasters,* 13; S. Suthawan and E. Barbier, "Valuing Mangrove Conservation in Southern Thailand," *Contemporary Economic Policy* 19 (2001): 109–122, 116.

20. Jane Spencer, "Forest Clearing May Have Worsened Toll," *Wall Street Journal,* May 9, 2008, A10.

21. Wells et al., *In the Front Line,* 21.

22. Sudmeier-Rieux et al., *Ecosystems, Livelihoods and Disasters,* 15.

23. Ibid., 35.

24. Paul Francis Diehl and Nils Petter Gleditsch, eds., *Environmental Conflict: An Anthology* (Boulder, Colo.: Westview Press, 2001).

25. Abramovitz, *Unnatural Disasters,* 16.

26. Ibid., 41.

27. Jared Diamond, *Collapse: How Societies Choose to Fail or Succeed* (New York: Penguin Books, 2005), 329–357; Sudmeier-Rieux et al., *Ecosystems, Livelihoods and Disasters,* 17.

28. Matthew L. Stutz and Orin H. Pilkey, "Global Distribution and Morphology of Deltaic Barrier Island Systems," *Journal of Coastal Research* 36 (2002): 694–707, 694–695; U.S. Geological Survey, *Post Hurricane Katrina Flights over Louisiana's Barrier Islands,* (Aug. 12, 2008) ww.usgs.gov (accessed July 3, 2008).

29. Stutz and Pilkey, "Global Distribution and Morphology," 694.

30. Ibid.

31. Ibid.

32. Orrin H. Pilkey and Mary Edna Fraser, *A Celebration of the World's Barrier Islands* (2003), 116; James G. Titus, "Greenhouse Effect, Sea Level Rise, and Barrier Islands: Case Study of Long Beach Island, New Jersey," *Coastal Management*, 18 (1990): 65–90, 68, fig. 3 (showing illustrations of Isle(s) Derniere(s) from 1890 to 1988); U.S. Geological Survey Marine and Coastal Geology Program, *Louisiana's Barrier Islands: A Vanishing Resource* (Nov. 3, 1995) marine.usgs.gov (accessed July 2, 2008).

33. Pilkey and Fraser, *Celebration of the World's Barrier Islands*, 130.

34. Wells et al., *In the Front Line*, 14–15.

35. Ibid.

36. Ibid.

37. "Global Warming Is Destroying Coral Reefs, Major Study Warns," *ScienceDaily*, Dec. 14, 2007, www.sciencedaily.com (accessed July 2, 2008).

38. Gautam Naik and Shai Oster, "Scientists Link China's Dam to Earthquake, Renewing Debate," Feb. 6, 2009, *Wall Street Journal*, online.wsj.com (accessed June 10, 2009) (reporting on paper published in the Chinese scientific journal *Geology and Seismology* in December 2008).

39. Abramowitz, *Unnatural Disasters*, 17 (citing research on deforestation in northwestern United States); Sudmeier-Rieux et al., *Ecosystems, Livelihoods and Disasters*, 12 (citing International Strategy for Disaster Reduction, *Living with Risk, A Global Review of Disaster Reduction Initiatives* (2004), www.unisdr.org, describing the Swiss experience); Karen Sudmeier-Rieux et al., *Disaster Risk, Livelihoods and Natural Barriers: Strengthening Decision-making Tools for Disaster Risk Reduction: A Case Study from Northern Pakistan* (World Conservation Union, 2007), 4 (linking landsides in the 2005 Pakistan earthquake to deforestation).

40. Andrew H. Baird, "Acehnese Reefs in the Wake of the Asian Tsunami," *Current Biology* 15 (2005): 1926–1930, 1926.

41. Sunil Raman, "Tsunami Villagers Give Thanks to Trees," *BBC News*, Feb. 16, 2005, news.bbc.co.uk (accessed July 2, 2008).

42. Alston Chase, *Playing God in Yellowstone: The Destruction of America's First National Park* (New York: Harvest Books 1987), 92–93.

43. Thomas P. Holmes, Jeffrey P. Prestemon, and Karen L. Abt, eds., *The Economics of Forest Disturbances* (New York: Springer, 2008).

44. American Lands Alliance, *National Forest Fire Policy Statement* (2000), fire-ecology .org (accessed July 5, 2008).

45. Sudmeier-Rieux et al., *Ecosystems, Livelihoods and Disasters*, 1.

46. David Sneed, "Officials: Health Risks in New Orleans Unclear," Tri-City Herald (Tri-Cities, Wash.), Sept. 29, 2005, D3.

47. Ibid.

48. Janet Abramovitz, *Unnatural Disasters*, 17–18.

49. Sudmeier-Rieux et al., *Ecosystems, Livelihoods and Disasters*, 1.

50. Sneed, "Officials."

51. John McQuaid and Mark Schleifstein, *Path of Destruction: The Devastation of New Orleans and the Coming Age of Superstorms* (New York: Little Brown, 2006), 68; Amy Wold and Mike Dunn, "Tests Find N.O. Water Less Toxic Than Feared," *Baton Rouge Advocate*, Oct. 5, 2005, 1B; "New Orleans Flood Water Testing Raises Concerns," *USA Today*, Oct. 11, 2005, Nation sec.; David Sneed, "Officials."

52. Sudmeier-Rieux et al., *Ecosystems, Livelihoods and Disasters*, 1.

53. Ibid., 18.

54. Ibid., 31.

55. Leslie Eaton, "A New Landfill in New Orleans Sets Off a Battle," *New York Times*, May 8, 2006, A1.

3. System Failures and Fairness Deficits

1. P. R. Tamrakar, *State of Forest Genetic Resources Conservation and Management in Nepal*, working paper (Rome: Forest Resources Development Service, Forest Resources Division, 1988) (citing Master Plan for the Forestry Sector. Ministry of Forest and Soil Conservation of Nepal, 1988).

2. David Zetland, *Killing the Golden Goose? Tourism and Deforestation in Nepal*, working paper (Berkeley: Department of Agricultural and Resource Economics, University of California, Berkeley, 2005) (noting that Nepal's "tourist load" adds 85 percent to the country's base fuel).

3. See Tamrakar, *State of Forest Genetic Resources*, 2, table 1. While deforestation and forest thinning are serious problems in Nepal, in some regions the problem does appear to have been unintentionally overstated. See Stan Stevens, "Tourism and Deforestation in the Mt. Everest Region of Nepal," *Geographic Journal* 169 (2003): 255 (arguing that the extent of deforestation in Sagarmartha [Mt. Everest] National Park in the Khumbu region has been overstated).

4. Zetland, "Killing the Golden Goose," 7.

5. Ibid., 3.

6. James Salzman, "Creating Markets for Ecosystem Services: Notes from the Field," 80 *New York University Law Review* 80 (2005): 870, 879–880.

7. Rapanos v. United States, 547 U.S. 715 (2006).

8. Ibid., 64. Other references to this colloquy may be found at 63–65.

9. Janet Abramovitz, *Unnatural Disasters*, ed. Linda Starke (Washington, D.C.: Worldwatch Institute, 2001), 42–45.

10. James Salzman, "A Field of Green? The Past and Future of Ecosystem Services," *Journal of Land Use and Environmental Law* 21 (2006): 133, 134.

11. Robert R. M. Verchick, "Feathers or Gold? A Civic Economics for Environmental Law," 25 *Harvard Environmental Law Review* 25 (2001): 95, 104–06.

12. David Brower, "Public Interest Environmental Law Conference," lecture, University of Oregon, Eugene, Mar. 4, 1999.

13. James Rasband et al., *Natural Resources Law and Policy* (New York: Foundation Press, 2004), 70–71.

14. Salzman, "Field of Green," 135.

15. Philippe Rekacewicz and Emanuelle Bournay, *Millennium Ecosystem Assessment* (Washington, D.C.: World Resources Institute, 2005), 9, fig. 8 (emphasis added).

16. Ibid., 49.

17. Ibid.

18. Salzman, "Field of Green," 135.

19. J. B. Ruhl, Steven Kraft, and Christopher Lant, *The Law and Policy of Ecosystem Services* (Washington, D.C.: Island Press, 2007).

20. William H. Rodgers, *Environmental Law*, 2nd ed. (St. Paul, Minn.: West, 1994), 36–39.

21. W. Edward Steinmueller, "Technological Infrastructure in Information Technology Industries," in *Technological Infrastructure Policy: An International Perspective*, ed. Morris Teubal et al. (New York: Springer, 1996), 117.

22. Salzman, "Field of Green," 135.

23. Ibid., 36.

24. Ibid.

25. Ibid., 35.

26. Gretchen Daily and Katherine Ellison, *The New Economy of Nature: The Quest to Make Conservation Profitable* (Washington, D.C.: Island Press, 2002), 13.

27. Salzman, "Field of Green," 146.

28. Daily and Ellison, *New Economy of Nature*, 231 (noting the important role of government in creating and regulating new kinds of markets and in setting substantive limits, as in the Endangered Species Act, to resource destruction).

29. David Zetland, "Killing the Golden Goose?"

30. Bruce Katz and Robert Puentes, "Clogged Arteries," *Atlantic Monthly*, March 2008, 38.

31. Ibid.

32. See, e.g., Matthew D. Adler, "Policy Analysis for Natural Hazards: Some Cautionary Lessons from Environmental Policy Analysis," *Duke Law Journal* 56 (2006): 1, 7; Richard A. Posner, *Economic Analysis of Law*, 5th ed. (Austin, Tex.: Aspen, 1998), 13. But see Louis Kaplow and Steven Shavell, *Fairness versus Welfare* (Cambridge, Mass.: Harvard University Press, 2002) (arguing that welfare should be the only controlling factor in legal analysis and that fairness, as a freestanding consideration, should be ignored).

33. Joseph William Singer, *Property Law: Rules, Policies, and Practices*, 4th ed. (New York: Aspen, 2006), xlix.

34. Ibid., 1 (quoting Jennifer Nedelsky, "Law, Boundaries, and the Bounded Self," *Representations* 30 [Spring 1990]: 162, 169, (internal quotation marks omitted).

35. Ibid.; see also Joseph William Singer, *Entitlement: The Paradoxes of Property* (New Haven, Conn.: Yale University Press, 2000).

36. Joseph William Singer, "Something Important in Humanity," *Harvard Civil Rights-Civil Liberties Law Review* 37 (2002): 103.

37. See, e.g., Isaiah Berlin, *Two Concepts of Liberty* (Oxford: Clarendon Press, 1958), 6–19.

38. Nancy Levit and Robert R. M. Verchick, *Feminist Legal Theory: A Primer* (New York: New York University Press, 2006), 226; Martha C. Nussbaum, *Women and Human Development: The Capabilities Approach* (New York: Cambridge University Press, 2000).

39. See Zetland, "Killing the Goose," 10.

40. For the argument that market-based prices already reflect fairness concerns, see Kaplow and Shavell, *Fairness versus Welfare*, 455–457.

4. Tending Our Gardens

1. *Sag wagon*: the van that follows bicyclists in races and organized rides to pick up riders no longer able to participate in the event.

2. The categories are a little subjective but will serve my purpose here. For other taxonomies of environmental protection strategies, see Jonathan Remy Nash, *Environmental Law and Policy: The Essentials* (Waltham, Mass.: Aspen, 2009); J. B. Ruhl, Steven E. Kraft, and Christopher L. Lant, *The Law and Policy of Ecosystem Services* (New York: Island Press, 2007), 87; James Salzman, "Creating Markets for Ecosystem Services: Notes from the Field," *New York University Law Review* 80 (2005): 870; Carol M. Rose, "Rethinking Environmental Controls: Management Strategies for Common Resources," *Duke Law Journal* 1 (1991).

3. For instance, the standards of nuisance law could be considered a kind of public regulation. Or, where the remedy is monetary, nuisance law could be considered part of a regime of economic incentives. Public trust law can be seen as a set of rules governing the management of public resources. Takings law, at least the part dealing with regulatory takings, can be seen as a limit on public regulation.

4. Ruhl, et al., *Ecosystem Services*, 113.

5. Alyson Flournoy, Margaret Clune Giblin, and Matt Schudtz, *Squandering Public Resources*, (2007), Center For Progressive Reform, www.progressivereform.com (accessed June 20, 2008), 2.

6. Ibid., 6–9.

7. Ruhl et al., *Ecosystem Services*, 114–115.

8. Conservation Act, U.S. Code 16 (1960), sec. 531.

9. See Ruhl et al., *Ecosystem Services*, 115.

10. Ibid.

11. Healthy Forest Restoration Act, 16 U.S.C. sec. 6501 (2004).

12. See U.S. Government Accountability Office, *Forest Service: Use of Categorical Exclusions for Vegetation Management Projects, Calendar Years 2003 through 2005*, GAO-07-99 (Oct. 2006), www.gao.gov (accessed September 18, 2008) (citing "National Environmental Protection Act Documentation Needed for Fire Management Activities; Categorical Exclusions," *Federal Register* 68 no. 33814 [June 2003]); "National Environmental Protection Act Documentation Needed for Limited Timber Harvest," *Federal Register* 68 no. 44598 [July 2003]); U.S. Department of Agriculture et al., "Administrative Actions to Implement the President's Healthy Forests Initiative," www.whitehouse.gov (accessed Sept. 18, 2008), 3–4.

13. Flournoy et al., *Squandering Resources,* 23 (citing Ross W. Gorte, "Wildfire Protection in the 108th Congress," *Congressional Service Report* no. RS22024 [Jan. 2005]).

14. Sierra Club, *De-bunking the "Healthy Forests Initiative,"* www.sierraclub.org (accessed July 28, 2008); see also Flournoy et al., *Squandering Resources,* 23; Jesse B. Davis, "The Healthy Forests Initiative: Unhealthy Policy Choices in Forest and Fire Management," comment, *Environmental Law* 34 (2004): 1209. Mike Leahy, "EMS Ignores True Nature of Planning," *Environmental Forum* (Jan./Feb. 2004): 31; Ross W. Gorte, "Wildfire Protection in the 108th Congress," *Congressional Research Service Report* no. RS22024 (January 2005); William W. Buzbee et al., *Regulatory Underkill: The Bush Administration's Insidious Dismantling of Public Health and Environmental Protections*, Center for Progressive Reform, www.progressivereform.org (accessed July 28, 2008).

15. Frank Ackerman and Lisa Heinzerling, *Priceless: On Knowing the Price of Everything and the Value of Nothing* (New York: New Press, 2005), 7.

16. Flournoy et al., *Squandering Resources,* 24.

17. Citizens for Better Forestry v. U.S. Dep't of Agric., 481 F.Supp. 2d 1059, 1100 (N.D. Cal. 2007).

18. Citizens for Better Forestry v. U.S. Department of Agriculture, 632 F. Supp. 2d (N.D. Cal., 2009).

19. Flournoy et al., *Squandering Resources,* 26–27 (citing U.S. Government Accountability Office, *Forest Service: Use of Categorical Exclusions for Vegetation Management Projects, Calendar Year 2003–2005*, GAO 07-99 (October 2006), 1–2).

20. Jeff Stein, Peter Moreno, David Conrad, and Steve Ellis, *Troubled Waters: Congress, the Corps of Engineers, and Wasteful Water Projects,* (Washington, D.C.: Taxpayers for Common Sense and National Wildlife Federation, March 2000).

21. U.S. Army Corps of Engineers, *The U.S. Army Corps of Engineers: A Brief History* (November 26, 2007), www.hq.usace.army.mil (accessed January 19, 2010).

22. John Barry, *Rising Tide: The Great Mississippi Flood of 1927 and How It Changed America* (New York: Simon and Schuster, 1998), 279.

23. The Federal Flood Control Act of 1928, 45 Stat. 534, as amended 33 U.S.C. sec. 701 et seq. (1994).

24. Flood Control Act of 1936, sec. 5.

25. William E. Leuchtenburg, *Flood Control Politics: The Connecticut River Valley Problem 1927—1950* (Cambridge, Mass.: Harvard University Press, 1953), 96–105; see also Joseph L. Arnold, author's note to *The Evolution of the 1936* Flood Control Act (Fort Belvoir, Va.: Army Corps of Engineers, 1988); Christine Klein and Sandra Zellmer, "Mississippi River Stories," *Southern Methodist University Law Review* 60 (2007): 1482–1488.

26. Oliver Houck, "Can We Save New Orleans," *Tulane Environmental Law Journal* 19 (2006): 12.

27. Ibid., 12.

28. Ibid.

29. Ibid., 8.

30. Arnold, *Evolution of the 1936 Flood Control Act,* 95; see also Klein and Zellmen, "Mississippi River Stories," 1536.

31. Russ Feingold, *On Reforming the Army Corps of Engineers, as Prepared for Delivery from the Senate Floor July 19, 2006,* www.feingold.senate.gov (accessed September 19, 2008).

32. Ibid.

33. Statement of Anu Mittal, Director National Resources and Environment, U.S. Government Accountability Office, Testimony before the Subcommittee on Government Reform, House of Representatives, *Corps of Engineers, Observations on Planning and Project Management Process for Civil Works Program,* GAO 06-529T, Mar. 15, 2006 (reporting on U.S. General Accounting Office, *Delaware River Deepening Project: Comprehensive Reanalysis Needed,* GAO-02-604 [Washington, D.C.: June 7, 2002], 26–27); U.S. General Accounting Office, *Oregon Inlet Jetty Project: Environmental and Economic Concerns Still Need to Be Resolved,* GAO-02-803 (Washington, D.C.: Sept. 30, 2002): 14; U.S. General Accounting Office, *Corps of Engineers: Improved Analysis of Costs and Benefits Needed for Sacramento Flood Protection Project,* GAO-04-30 (Washington, D.C.: Oct. 27, 2003), 1; U.S. General Accounting Office, *Corps of Engineers: Effects of Restrictions on Corps' Hopper Dredges Should Be Comprehensively Analyzed,* GAO-03-382 (Washington, D.C.: Mar. 31, 2003), 1; U.S. Government Accountability Office, *Army Corps of Engineers: Improved Planning and Financial Management Should Replace Reliance on Reprogramming Actions to Manage Funds,* GAO-05-946, Sept. 16, 2005, 1.

34. Ibid.

35. Russ Feingold, "On Reforming the Army Corps of Engineers."

36. Daniel B. Wood, "California Mired in Its Own Levee Crisis," *Christian Science Monitor,* Mar. 21, 2006, www.csmonitor.com (accessed June 10, 2009) (describing testimony from the California Department of Water Resources).

37. Andrew Bridges, "Development Raises Flood Risk across U.S.," Associated Press, Feb. 18, 2006, news.yahoo.com (accessed July 7, 2008). Under a nationwide levee review, mandated by Congress after Katrina, the U.S. Army Corps of Engineers recently released a list of 122 levees in 30 states and territories that it says have an unacceptable risk of failing. *U.S. Army Corps of Engineers, Levees of Maintenance Concern,* (Feb. 1, 2007), www.hq.usace.army.mil (accessed July 7, 2008). Thirty-eight are located in California, thirty-six in the Sacramento District, one in the Los Angeles District, and one in the San Francisco District; other jurisdictions with multiple listings are Alaska, Arkansas, Connecticut, the District of Columbia, Hawai'i, Idaho, Illinois, Kentucky, Maryland, Massachusetts, Michigan, New Mexico, Oregon, Pennsylvania, Puerto Rico, Rhode Island, and Washington. Ibid.

38. John McQuaid, "Never Again? The Politics of Preventing another Katrina," *Mother Jones,* Aug. 29, 2007, www.motherjones.com (accessed September 22, 2008).

39. Ibid.

40. Louisiana Coastal Protection and Restoration Authority (LCPRA), *Integrated Ecosystem Restoration and Hurricane Protection: Louisiana's Comprehensive Master Plan for a Sustainable Coast* (2007), www.lacpra.org (accessed July 31, 2008), 103.

41. LCPRA, "Coastal Protection and Restoration Authority Approves State Coastal Master Plan," press release, April 2007, www.lacpra.org (accessed September 17, 2008).

42. LCPRA, *Integrated Ecosystem Restoration,* executive summary.

43. Ibid.

44. Ibid.

45. Ibid.

46. Ibid.

47. Ibid.

48. Ibid.

49. Ibid.

50. Adaptive management is gradually being introduced to many areas of resource management. For more on this trend, see J. B. Ruhl, *Regulation by Adaptive Management—Is It Possible?* Florida State University College of Law, Law and Economics Paper no. 05-19, Social Science Research Network, www.ssrn.com (accessed January 19, 2010); Dan Tarlock, "Slouching toward Eden: The Eco-pragmatic Challenges of Ecosystem Revival," *Minnesota Law Review* 87 (2003): 1173; Judy L. Meyer, "The Dance of Nature: New Concepts in Ecology," *Chicago Kent Law Review* 69 (1994): 875; Holly Doremus and W. Michael Hanemann, "The Challenge of Dynamic Water Management," *UCLA Journal of Environmental Law and Policy* 26 (2008) 55–75.

51. See, generally, LCPRA, *Integrated Ecosystem Restoration,* app. C (summarizing comments made at public meetings).

52. Multiple Lines of Defense Team, *Comments on the Draft of Integrated Ecosystem Restoration and Hurricane Protection: Louisiana's Comprehensive Master Plan for Sustainable Coast* (April 2, 2007), www.mlods.org (accessed Sept. 22, 2008), 15, 18.

53. Multiple Lines of Defense Team, *Comprehensive Recommendations Supporting the Use of the Multiple Lines of Defense Strategy to Sustain Coastal Louisiana* (Aug. 17, 2007), www.mlods.org (accessed Sept. 22, 2008), 101, fig. 52.

54. Ibid., 106.

55. The actual quotation, attributed to Derek Bok, is "If you think education is expensive, try ignorance." See Derek C. Bok, www.quoteworld.org (accessed Sept. 22, 2008).

56. Louisiana Coastal Wetlands and Restoration Task Force, *Coastal Wetlands Planning, Protection and Restoration Act: A Response to Louisiana's Land Loss,* (Apr. 17, 2007), www.lacoast.gov, 10.

57. Kenneth J. Bagstad, "Taxes, Subsidies, and Insurance as Drivers of Coastal Development," *Ecological Economics* 63 (2007): 285–288.

58. White House Council on Environmental Quality, Proposed National Objectives, Principles and Standards for Water and Related Resources Implementation Studies, Dec. 3, 2009, pp. 1, 6, available at www.whitehouse.gov (accessed Jan. 13, 2010).

59. Ruhl et al., *Ecosystem Services,* 140 (quoting Virginia C. Veltman, "Banking on the Future of Wetlands Using Federal Law," *Northwestern University Law Review* 89 [1995]: 673,674.

60. Ibid., 141.

61. P. J. Huffstutter, "Keeping Pace with Iowa's Floods," *Los Angeles Times,* June 16, 2008, Main News sec.

62. Joel Achenbach, "Iowa Flooding Could Be an Act of Man, Experts Say," *Washington Post,* June 19, 2008, A1.

63. See ibid.

64. *Coastal Zone Management Act* National Oceanic and Atmospheric Adminstration, *see* About Coastal Zone Management Act (August 23, 2007), www.coastalmanagment.noaa.gov (accessed September 24, 2008).

65. Ruhl et al., *Ecosystem Services,* 144.

66. Coastal Zone Management, U.S. Code 16 U.S.C. sec. 1455(d)(9) (1972); ibid., 144.

67. Ruhl et al., *Ecosystem Services,* 145.

68. Coastal Wetlands, N.J. *Statutes Annotated,* sec. 13:9A-1.a (West 2003); Ibid.

69. Ruhl et al., *Ecosystem Services,* 145 (quoting Bureau of Beaches and Coastal Systems Rules and Procedures, *Florida Administrative Code Annotated,* secs. 62B-33.005[4]–[5] [West 2007]).

70. Ibid., 145 (quoting Mass. Regulations Code title 301, sec. 21.98[6] [West 2007]). (internal quotation marks omitted).

71. Ibid., 275; Salzman, "Creating Markets," 112.

72. See James Rasband et al., *Natural Resources Law and Policy* (New York: Foundation Press, 2004), 119–127, 136–137, 1231–1237, 1033–1037, 940–924 (discussing subsidies for agriculture, land development, and logging, as well as for mining and grazing); Mark Clayton, "U.S. House Takes on Big Oil," *Christian Science Monitor*, Jan. 18, 2007 (discussing subsidies for oil and gas industry), www.csmonitor.com (accessed Jan. 16, 2009).

73. National Forest Protection Alliance, *National Forest Fact Sheet*, (undated, www.rso.cornell.edu (accessed August 2, 2008).

74. Tax Payers for Common Sense, "Timber Subsidies Reach Record Level, According to New Report," press release, Mar. 6, 2001, www.commondreams.org (accessed September 22, 2008).

75. See Scott M. Swinton et al., "Ecosystem Services from Agriculture: Looking beyond the Usual Suspects," *American Journal of Agricultural Economics* 88 (2006): 1160–1166.

76. Millennium Ecosystem Assessment, *Ecosystems and Human Well-Being: Synthesis* (Washington D.C.: Island Press, 2005), 96,

77. Ruhl et al., *Ecosystem Services*, 277.

78. "Are You Being Served?" *Economist*, Apr. 23, 2005, 76.

79. Ibid.

80. Surface Mining Control and Reclamation Act, 30 U.S.C. secs. 1234–1328 (1977); U.S. Department of Interior, Office of Surface Mining, *Surface Coal Mining Reclamation: 25 Years of Progress, 1977–2002* (Jan. 14, 2003), www.osmre.gov (accessed Aug. 2, 2008).

81. See Comprehensive Environmental Response, Compensation and Liability Act (also known as Superfund), U.S. Code 42 (1988) secs. 9601 et seq. The tax, which generated around $1.5 billion each year, was in place from 1980, when the statute took effect, until 1995, when Congress allowed the tax to expire. During that time, the trust fund of money available to address so-called orphan sites was funded 80 percent by the tax and 20 percent by general federal revenue. By 2004, nine years after the tax expired, the trust fund was relying entirely on general federal revenue. Thus one could say that the trust fund went from a mainly "polluter pays" program to a "beneficiary pays" program. See generally, Rena Steinzor and Margaret Clune, *The Toll of Superfund Neglect: Toxic Waste Dumps and Communities at Risk* (2006), Center for Progressive Reform, www.progressivereform.org (accessed Aug. 15, 2008).

82. Domenici-Landrieu Gulf of Mexico Energy Security Act of 2006, S.3711, Senate, 109th Cong., 2d sess. (2006).

83. Louisiana Coastal Protection and Restoration Authority, *The Coastal Protection and Restoration Fund*, www.lacpra.org (accessed Sept. 22, 2008).

84. Ibid.

85. In 1949, Governor Earl Long was offered a deal by President Truman to accept a percentage of offshore oil revenue and turned it down. Louisiana was left with nothing. Governor Earl's poor judgment cost Louisiana billions.

86. Rasband et al., *Natural Resources*, 980–984.

87. America's Wetland Foundation, *America's Wetland*, www.americaswetland.com (accessed Aug. 12, 2008); Daniel Zwerdling, "Who Pays to Repair Louisiana's Wetlands?" *Weekend Edition Saturday*, National Public Radio, May 6, 2006.

88. America's Wetland Foundation, *America's Wetland*.

89. Zwerdling, "Who Pays to Repair Louisiana's Wetlands?"

90. See Coast 2050, "WRDA Funding: A Crucial Step in Saving Louisiana's Coast, *Watermarks* 19 (2001): 3.

91. Ibid.

92. The takings clause states: "nor shall private property be taken for public use, without just compensation," U.S. Constitution, Fifth Amendment.

93. See Lucas v. South Carolina Coastal Council, 505 U.S. 1003 (1992) (a regulation that deprives an owner of all "economically viable use of his land" constitutes a taking, unless the regulations simply makes explicit what already inheres in the title itself, by way of background principles of nuisance and other property law); City of Tigard v. Dolan, 512 U.S. 374 (1994) (the government may not require a person to give up a property right in exchange for a discretionary development permit where the property right sought has little or no relationship to the benefit).

94. See Tahoe-Sierra Preservation Council v. Tahoe Regional Planning Agency, 535 U.S. 302 (2002); Richard J. Lazarus, "Lucas Unspun," *Southeastern Environmental Law Journal* 16 (2007): 13.

95. John G. Sprankling, "The Antiwilderness Bias in American Property Law," *Chicago Law Review* 63 (1996): 519.

96. Sprankling, "Antiwilderness Bias," 554.

95. Ibid., 569.

96. Ruhl, Kraft, and Lant, *Ecosystem Services*, 109; see also J. B. Ruhl, "Toward a Common Law of Ecosystem Services," *St. Thomas Law Review* 18 (2005): 1, 6–7, 12–13 (noting the promise of innovation in the common law but describing its failure to establish rights to ecosystem services).

99. I should make it clear that the common law has sometimes contributed in important ways to environmental protection. Before the boom of federal environmental statutes in the 1970s, nuisance and trespass suits were one of the most successful tools in fighting air and water pollution. Even today, lawsuits based on common

law play an important role in pollution protection. Indeed, such lawsuits are one of the few ways that plaintiffs can recover monetary damages to compensate them for harms suffered. Yet despite this success, it must be recognized that common law doctrines have not lived up to their potential, and this is mainly because Americans courts have historically been reluctant to apply doctrinal principles to their logical conclusions in environmental cases. As I argue later in this chapter, common law doctrines like nuisance and the related public trust doctrine could be reinvigorated by courts to further strengthen environmental protection. For a more complete discussion of how the common law might be turned into a more powerful environmental tool, see Alexandra B. Klass, "Common Law and Federalism in the Age of the Regulatory State," *Iowa Law Review*, 92 (2007): 545.

100. See, e.g., Southwest Weather Research, Inc. v. Rounsaville, 320 S.W.2d 211 (Tex. Civ. App. 1958).

101. Prah v. Maretti, 321 N.W.2d 182, 189 (Wis. 1982); Tenn v. 889 Associates, Ltd., 500 A.2d 366, 369 (N.H. 1985) citing *Prah v. Maretti*.

102. Joseph William Singer, *Introduction to Property*, 2nd ed. (New York: Aspen, 2005), sec. 3.5.1–3.5.2 (describing rights to lateral and subjacent support).

103. See, e.g., Klausmeyer v. Makaha Valley Farms, 41 Haw. 287, 1956 WL 10310 (1956) (owner is liable for removal of beach sand on his property where such removal, combined with the foreseeable forces of wind and waves, depletes sand on adjoining property thereby causing damage); Murray v. Pannaci, 64 N.J. Eq. 147, 153, 53 A. 595 (1902) ("a person who excavates on his land in such a manner as to let in the sea, which undermines and injures the adjoining land of another, is liable for the injury so placed"); see also Attorney-General v. Tomline, 14 Ch. Div. 58 (1880) (holding, under English private law, that the owner of property in the intertidal zone may not remove pebbles from that property so as to expose neighboring land to the inroads of the sea). Barasich v. Columbia Gulf Transmission Co., 467 F.Supp. 2d 676 (E.D. La. 2006).

104. Ibid.

105. Ibid., 678.

106. Ibid., 680.

107. Susan Pagano, Drew Douglas, and Maybelle Cagle, Bureau of National Affairs, "Oil, Gas, Pipeline Companies Sued for Damages over Hurricane Katrina," *Class Action Litigation Report* 6, no. 19 (Oct. 14, 2005), www.litigationcenter.bna.com (accessed Jan. 12, 2009).

108. Ibid.

109. *Barasich*, 467 F.Supp. 2d at 695.

110. Ibid., 690.

111. Restatement (Second) on Torts, sec 821B.

112. *Barasich*, 467 F.Supp. 2d at 694 (emphasis added).
113. W. Page Keeton et al., *Prosser and Keeton on the Law of Torts*, 5th ed. (St. Paul, Minn.: West, 1984), sec. 88B (citing case law from Idaho, Iowa, Maryland, Minnesota, and New York).
114. Ibid., 695.
115. See General Accounting Office Reports and Testimony, *Alaska Native Villages: Villages Affected by Flooding and Erosion Have Difficulty Qualifying for Federal Assistance*, June 20, 2004, www.gao.gov (accessed July 2, 2008); Emma Schwartz, "Attacking Climate Change in Court," *U.S. News and World Report*, Mar. 13, 2008, 12; Complaint for Damages, Native Village of Kivalina v. ExxonMobil et al., CV 08-1138 (N.D. Calif., Filed Feb. 26, 2008).
116. Native Village of Kivalina v. ExxonMobil Corporation, 663F. Supp. 2d 863 (N.D. Cal., 2009).
117. The two cases also raise questions of justiciability, that is, a court's ability to exercise legitimate authority over the matter and to provide an adequate resolution to the dispute. See *Barasich*, 467 F. Supp. 2d at 680–681 (finding that plaintiffs' claims were justifiable despite their broad scope); Kivalina, 663 F Supp. 2d 863.
118. National Audubon Society v. Superior Court, 189 Cal. Rptr. 346 (Cal. 1983); see also Mark Dowie, "In Law We Trust: Can Environmental Legislation Still Protect the Commons?" *Orion Online*, July/August 2005 www.oriononline.org (accessed Jan. 19, 2010).
119. Magna Carta, English Bill of Rights (1225); see also Alexandra B. Klass, "Modern Public Trust Principles: Recognizing Rights and Integrating Standards," *Notre Dame Law Review* 82 (2006): 699.
120. 146 U.S. 387 (1892).
121. Ibid., 453.
122. Robin Kundis Craig, "Comparative Guide to the Eastern Public Trust Doctrines: Classifications of States, Property Rights, and State Summaries," *Penn State Environmental Law Review* 16 (2007): 7.
123. Joseph L. Sax, "The Public Trust Doctrine in Natural Resource Law: Effective Judicial Intervention," *Michigan Law Review* 68 (1970): 473.
124. Craig, "Comparative Guide," 11.
125. Ibid., 11.
126. Ibid., 29–30.
127. Ibid., 1.
128. Ibid., 25.
129. Ibid., 22 (quoting Cinque Bambini Partnership v. State, 491 So.2d 508, 512 [Miss. 1986]).
130. Sierra Club v. Kiawah Resort Assocs., 456 S.E.2d 397, 402 (S.C. 1995), quoting Gregg L. Spyridon and Sam A. Leblanc, "The Overriding Public Trust in Privately

Owned Natural Resources: Fashioning a Cause of Action," 6 *Tulane Environmental Law Journal* 6 (1993): 287, 291.

131. Louisiana Constitution, art. 9, sec. 1.

132. Avenal v. State, 886 So.2d 1085,1101 (La. 2004) quoting Save Ourselves, Inc. v. Louisiana Environmental Control Com'n, 452 So. 2d 1152, 1157 (La. 1984).

133. J. B. Ruhl and James Salzman, "Ecosystem Services and the Public Trust Doctrine: Working Change from Within," *Southeastern Environmental Law Journal* 15 (2006). 223.

134. "Huge Freshwater Diversion Project Rescues Louisiana Wetlands," *Public Works*, Mar. 1992, 44; Louisiana Department of Natural Resources, "Caernarvon Project Given an 'A' by Coastal Specialists," news release, Oct. 27, 1997.

135. Louisiana Dep't of Natural Resources and U.S. Army Corps of Engineers, *Caernarvon Freshwater Diversion Project* (March 11, 1998), www.lacoast.gov.

136. Aaron Kuriloff, "Justice on the Half Shell: Louisiana's Oyster Farmers Get a Raw Deal in Court," *Legal Affairs*, May/June 2005, www.legalaffairs.org (accessed Sept. 22, 2008).

137. *Avenal*, 886 So.2d at 1097 n.14, quoting the Coastal Wetlands Restoration Advisory Clause.

138. Louisiana Revised Statute Ann. sec. 56:425(C)

139. *Avenal*, 886 So.2d at 1091, n. 8 (Weimer, J., concurring).

140. Ibid.

141. Alexandra Klass, "Modern Public Trust Principles: Recognizing Rights and Integrating Standards," *Notre Dame Law Review* 82 (2006): 699, 743.

142. Parker v. New Hanover County, 619 S.E.2d 868 (N.C. Ct. App. 2005) described in ibid., 736–737.

143. *Parker*, 619 S.E.2d at 875–876.

144. Ibid.

145. Lake Bistineau Pres. Soc'y, Inc. v. Wildlife and Fisheries Comm'n, 895 So.2d 821, 827 (La. Ct. App. 2005), cert. denied, 544 U.S. 1044 (2005); In re Water Use Permit Applications, 9 P.3d 409, 425–430 (Haw. 2000); Weden v. San Juan County, 958 P.2d 273, 283–284 (Wash. 1998).

146. Klass, "Modern Public Trust Principles," 738–739.

147. 201 N.W.2d 761 (Wis. 1972).

148. *Just*, 201 N.W.2d at 768.

149. *Tenn*, 500 A.2d at 327.

150. Josh Waitzkin, *The Art of Learning: A Journey in the Pursuit of Excellence* (New York: Free Press, 2007), 74 (eight-time national chess champion explaining the contrasting ways that expert players and less expert players evaluate their positions in a chess game); see, generally, Josh Waitzkin, *Attacking Chess: Aggressive Strategies and Inside Moves from the U.S. Junior Chess Champion* (New York: Fireside, 1995).

5. Backwater Blues

Epigraph. Al Young, "Watsonville after the Quake," in *Poems: "Something about the Blues"* (Naperville, Ill.: Sourcebook Media Fusion, 2007).

1. William Faulkner, "Old Man," in *The Portable Faulkner*, ed. Malcolm Cowley (New York: Penguin Books, 1977), 481, 486.

2. While often presented as a self-contained work, this long story first appeared as part of Faulkner's 1939 novel *Wild Palms*. See 479–480 (editor's note).

3. Ibid., 479–480.

4. Charley Patton, *High Water Everywhere Pt. 1/High Water Everywhere Pt. 2*, Paramount 12909, April 1930, 78 rpm. Ethnomusicologist David Evans estimates there were "25 to 30 records by blues artists on or related to the 1927 flood"; Bessie Smith's "Back Water Blues" may be the most well-known blues tune associated with that flood; ironically, Smith recorded the song in February 1927—after the heavy rains had begun but *before* the river actually overflowed. Evans, *Fatal Flood: Delta Blues Music* (1999–2000), www.pbs.org (accessed July 8, 2008). See Bessie Smith, "Back Water Blues," *The Essence of Bessie Smith*, mBopGlobal-Capriccio, 2006.

5. Led Zeppelin, "When the Levee Breaks," *Led Zeppelin IV*, Atlantic, 1971, (music and lyrics credited to Jimmy Page, Robert Plant, John Paul Jones, John Bonham, and Memphis Minnie). The piece is a retooling of the song first performed by Kansas Joe and Memphis Minnie, with nearly identical lyrics. See Kansas Joe and Memphis Minnie, "When the Levee Breaks," *Blues Classics by Memphis Minnie*, Columbia Records, 1929.

6. See Zora Neale Hurston, *Their Eyes Were Watching God* (1937) (Urbana: University of Illinois Press, 1978).

7. See John Steinbeck, *The Grapes of Wrath* (New York: Viking Press, 1939).

8. Although presented mainly as natural hazards, both the Okechobee floods and the Dust Bowl storms bore aspects of human contribution. The flood damage, for instance, was magnified by earlier decisions to drain the swamps around Lake Okechobee (destroying the area's natural capacity for absorbing floodwater) and to convert it to vegetable fields. Michael Grunwald, *The Swamp: The Everglades, Florida, and the Politics of Paradise* (Simon and Schuster, 2006). The dust storms of the 1930s, which included the abundant loss of topsoil, were intensified by unsustainable farming techniques that, in turn, were fueled by oppressive lending practices dictated by large eastern banks. Robert R. M. Verchick, "Steinbeck's Holism: Science, Literature, and Environmental Law," *Stanford Environmental Law Journal* 22 (2003): 1, 21.

9. Philip L. Fradkin, *The Great Earthquake and Firestorm of 1906: How San Francisco Nearly Destroyed Itself* (Berkeley: University of California Press, 2005), 211. Fradkin argues that accounts of this sentiment were exaggerated by outside observers.

10. Ted Steinberg, *Acts of God: The Unnatural History of Natural Disaster in America* (New York: Oxford University Press, 2000), 43–44 (describing research by social historian Gladys Hansen); Lawrence J. Vale and Thomas J. Campanella, "Introduction: The Cities Rise Again," in *The Resilient City: How Modern Cities Recover from Disaster*, ed. Vale and Campanella (New York: Oxford University Press, 2005), 3, 10; Fradkin, *Great Earthquake and Firestorm*, 3–4; Simon Winchester, *Crack in the Edge of the World: America and the Great California Earthquake of 1906* (New York: HarperCollins, 2005), 301–303. In contrast, the New Orleans Flood of 2005 inundated 80 percent of the city and resulted in more than 800 fatalities. Louisiana Department of Health and Hospitals, *Hurricane Katrina: Reports of Missing and Deceased*, August 2, 2006, www.dhh.louisiana.gov (accessed Sept. 22, 2008).

11. Steinberg, *Acts of God*, 44.

12. Fradkin, *Great Earthquake and Firestorms*, 211.

13. Ibid., 292.

14. Ibid., 295; Winchester, *Crack in the Edge of the World*, 330–331.

15. Daniel A. Farber, "Disaster Law and Inequality," *Law and Inequality* 25 (Summer 2007): 297, 298 (citing Winchester, *Crack in the Edge of the World*, 301).

16. "Gigantic Is This Project," *Los Angeles Times*, May 4, 1906, www.sfgenealogy.com (accessed 30 June 2008), 1. This news article, along with others related to the 1906 earthquake and its effects on women, can be found at the web site "San Francisco 1906 Earthquake Marriage Project," maintained by sfgenealogy.com.

17. Ibid. (quoting "Mrs. Jefferson D. Gibbs," president of the Woman's Parliament of Southern California).

18. Andrea Davies Henderson, "Disastrous Opportunities: Gender and Class Relations in San Francisco Relief Camps after the 1906 Earthquake and Fire," unpublished manuscript.

19. Faulkner, "Old Man," 486; see also John M. Barry, *Rising Tide: The Great Mississippi Flood of 1927 and How It Changed America* (New York: Touchstone, 1997), 122–123.

20. Barry, *Rising Tide*, 308.

21. Farber, "Disaster and Inequality," 300.

22. Ibid. (citing Barry, *Rising Tide*, 311).

23. Barry, *Rising Tide*, 417.

24. Ibid., 258; Judd Slivka, "Another Flood That Stunned America," *U.S. News and World Report* 139, no. 9 (2005): 26.

25. Grunwald, *Swamp*, 193.

26. Eliot Kleinberg, "The Florida Flood That Accounted for the Most Deaths of Black People in a Single Day (Until Katrina)," History News Network, Sept. 6, 2005, hnn.us (accessed June 30, 2008). Despite the title of Kleinberg's article, if Grunwald's estimate of black fatalities is correct, fewer African Americans per-

ished during Hurricane Katrina and the New Orleans flood than during Florida's 1928 hurricane.

27. Hurston, *Their Eyes Were Watching God,* 239.

28. Grunwald, *Swamp,* 192.

29. Kleinberg, *Florida Flood.*

30. Hurston, *Their Eyes Were Watching God,* 243.

31. Grunwald, *Swamp,* 194.

32. Kleinberg, *Florida Flood.*

33. Carl Nolte, "After the Fall," *San Francisco Chronicle,* Oct. 17, 1999, www.sfgate .com (accessed Aug. 4, 2008), SC-1.

34. Steinberg, *Acts of God,* 44–45.

35. Ibid.

36. Ibid.

37. William V. Flores, "Mujeres en Huelga, Cultural Citizenship and Gender Empowerment in a Cannery Strike," 210–254, 215, in *Latino Cultural Citizenship: Claiming Identity, Space, and Rights,* ed. William V. Flores and Rina Benmayor (Boston: Beacon Press, 2004); see also Mary C. Comerio, *Housing Repair and Reconstruction* (Dec. 9, 1997), National Information Service for Earthquake Engineering, nisee.berkeley.edu (accessed on June 30, 2008); Kathleen J. Tierney et al., eds., *Facing the Unexpected: Disaster Preparedness and Response in the United States* (Washington, D.C.: Joseph Henry Press, 2001), 118.

38. Young, "Watsonville." Readers interested in knowing more about Al Young should consult his official web site, Al Young: Poet Laureate of California, alyoung.org (accessed July 3, 2008).

39. Piers Blaikie et al., *At Risk: Natural Hazards, People's Vulnerability, and Disasters* (London: Routledge, 1994), 209.

40. See Matthew E. Kahn, "The Death Toll from Natural Disasters: The Role of Income, Geography, and Institutions," *Review of Economics and Statistics* 87, no. 2 (2005): 271–284, 271.

41. Ibid., i.

42. Ibid., 5.

43. Ibid., 3–4.

44. Seth Mydans, "Myanmar Rulers Still Impeding Access," *New York Times,* June 3 2008, A8; "Junta Is Stealing Aid, Relief Groups Asserts," *International Herald Tribune,* May 14, 2008, 1 (noting the Red Cross was estimating the final death toll to be in the range of "68,833 to 127,990").

45. Sai Soe Win Latt, "Cyclone Nargis Has Never Been 'Natural,'" opinion, *Irrawaddy* (Chiang Mai), May 2, 2008, www.irrawaddy.org (accessed May 28, 2008); "India, Thailand Warned Burma of Approaching Cyclone," *Irrawaddy* (Chiang Mai), May 8, 2008, www.irrawaddy.org (accessed May 28, 2008).

46. "Burmese Endure in Spite of Junta, Aid Workers Say," *New York Times*, June 18, 2008, A1.

47. Peace Way Foundation, *Burma Issues*, www.burmaissues.org (accessed June 2, 2008).

48. "Cyclone Survivors Victimized by Myanmar Soldiers," Associated Press, May 28, 2008, in.ibtimes.com (accessed May 29, 2008); see also Saw Yan Naing, "Cyclone Survivors Forcibly Evicted," *Irrawaddy* (Chiang Mai), May 24, 2008, www.irrawaddy .org (accessed May 28, 2008); Biswajyoti Das, "Cyclone Increases Army Looting on Burma Borders," *Irrawaddy* (Chiang Mai), May 24, 2008, www.irrawaddy.org (accessed May 28, 2008); Saw Yan Naing, "Reconstruction Just Propaganda, Say Rangoon Residents," *Irrawaddy* (Chiang Mai), May 20, 2008, www.irrawaddy.org (accessed May 28, 2008).

49. Chris Johnson, "Are Burma's Aid Delays Discriminatory?" *Christian Science Monitor*, May 14, 2008, www.csmonitor.comwww csmonitor.com (accessed January 11, 2009).

50. Ko Ko Maung and Saya San, "Why the Burmese Junta Failed to Respond to Cyclone Nargis," opinion, *Irrawaddy* (Chiang Mai), May 24, 2008, www .irrawaddy.org (accessed Jan. 11, 2009); Johnson, "Are Burma's Aid Delays Discriminatory?"

51. "Myanmar Starts Mass Evictions from Cyclone Camps," Reuters, May 30, 2008, www.reuters.com (accessed June 3, 2008); "Burma in the Aftermath of Cyclone Nargis: Death, Displacement, and Humanitarian Aid," statement by Scot Marciel, Ambassador to ASEAN and Deputy Assistant Secretary for East Asian and Pacific Affairs, before the Subcommittee on Asia, the Pacific, and the Global Environment of the House Committee on Foreign Affairs, Washington, D.C., May 20, 2008, www.state.gov (accessed June 3, 2008); Dan Nicholson, "Burma's Displaced People a Long-term Problem," opinion, *Irrawaddy* (Chiang Mai), May 14, 2008, www.irrawaddy.org (accessed June 3, 2008).

52. See Saw Yan Naing, "Children of the Cyclone," *Irrawaddy* (Chiang Mai), May 21, 2008, www.irrawaddy.org (accessed May 28, 2008); World Vision, "World Vision Opens Centers for Cyclone-affected Children," press release, May 14, 2008, www.reliefweb.int (accessed May 28, 2008); Violet Cho, "Many Cyclone Survivors Traumatized," *Irrawaddy* (Chiang Mai), May 13, 2008, www.irrawaddy.org (accessed May 28, 2008).

53. Alan Brown, "Burma Cyclone: British Aid Reaching Rangoon," *London Telegraph*, May 13, 2008, 15; see also U.S. Central Intelligence Agency, *World Fact Book*, *Burma* (last updated May 15, 2008), www.cia.gov (accessed May 28, 2008) (describing Myanmar as a source country for human trafficking).

54. Audra Ang, "Olympic Torch Reaches Earthquake-ravaged Sichuan Province," Associated Press, Aug. 3, 2008, www.chicagotribune.com (accessed Aug. 4, 2008);

Jill Drew, "For Survivors of Sichuan Quake, the Hard Lessons of Starting Over," *Washington Post*, July 21, 2008, A7.

55. Craig Simon, "After Earthquake, China's Poor Fear for the Future," Cox News Service, May 19, 2008, www.coxwashington.com (accessed May 28, 2008).

56. Drew, "For Survivors."

57. Loretta Chao et al., "China Earthquake Exposes a Widening Wealth Gap," *Wall Street Journal*, May 14, 2008, A1.

58. Ibid.

59. Andrew Jacobs, "Parents' Grief Turns to Rage at Chinese Officials," *New York Times*, May 28, 2008, A1.

60. Edward Wong, "China Presses Grieving Parents to Take Hush Money on Quake," *New York Times*, July 23, 2008, A1.

61. Hurston, *Their Eyes Were Watching God*, 215.

62. Young, "Watsonville."

63. Robert R. M. Verchick, "Feminist Theory and Environmental Justice," in *New Perspectives on Environmental Justice: Gender, Sexuality and Activism*, ed. Rachel Stein (New Brunswick, N.J.: Rutgers University Press, 2004)

64. For a firsthand account of Wangari Maathai's work and the founding of the Greenbelt Movement, readers should consult her riveting memoir, *Unbowed: A Memoir* (New York: Knopf, 2006).

65. Manual Pastor et al., "In the Wake of the Storm: Environment, Disaster, and Race after Katrina" (2006), www.russellsage.org (accessed May 28, 2008), 8.

66. See Manual Pastor et al., "Tainted Justice at the EPA," *Los Angeles Times*, Aug. 18, 2005, B13; Robert D. Bullard, "Unequal Environmental Protection: Incorporating Environmental Justice in Decision Making," in *Worst Things First? The Debate over Risk-based National Environmental Priorities*, ed. Adam M. Finkel and Dominic Golding (Washington, D.C.: Resources for the Future, 1994), 237–266.

67. See, e.g., Eileen Guana, et al., *Environmental Justice* (2005), Center for Progressive Reform, www.progressivereform.org (accessed Sept. 5, 2008) (citing many reports and studies).

68. Commission for Racial Justice of the United Church of Christ, *Toxic Wastes and Race in the United States: A National Report on the Racial and Socio-economic Characteristics of Communities with Hazardous Waste Sites*, (Cleveland: United Church of Christ, 1987), xiii.

69. Robert D. Bullard et al., "Toxic Wastes and Race at Twenty: 1987–2007 (2007), United Church of Christ, www.ucc.org (accessed Dec. 1, 2008), xii.

70. Pastor et al., *Wake of the Storm*, 8–9 (citing Rachel Morello-Frosch and Bill Jesdale, "Separate and Unequal: Residential Segregation and Estimated Cancer Risks Associated with Ambient Air Toxics in US Metropolitan Areas," *Environmental Health Perspectives* 114, no. 3 (2006): 386–393).

71. National Black Environmental Justice Network, *Environmental and Economic Injustice: Fact Sheet*, www.ejrc.cau.edu (accessed Dec. 1, 2008).

72. American Lung Association, *Lung Disease Data in Culturally Diverse Communities* (Washington, D.C.: American Lung Association 2005), 39.

73. Physicians for Social Responsibility—Los Angeles, *Childhood Lead Poisoning Prevention Act of 2007*, actionnetwork.org (accessed Nov. 23, 2008).

74. Ibid.

75. Marianne Lavelle and Marcia Coyle, "Unequal Protection," *National Law Journal*, Sept. 21, 1992, 1-2.

76. National Black Environmental Justice Network, "Environmental and Economic Injustice."

77. Robert Garcia et al., "Anatomy of the Urban Parks Movement: Equal Justice, Democracy, and Livability in Los Angeles," in *The Quest For Environmental Justice: Human Rights and the Politics of Pollution*, ed. Robert Bullard, et al. (San Francisco: Sierra Club Books, 2005), 149.

78. Samara F. Swanston, "Environmental Justice and Environmental Benefits: The Oldest, Most Pernicious Struggle and Hope for Burdened Communities," *Vermont Law Review* 23 (1999): 545, 557 (citing New York City Dep't of Planning, *New York City Comprehensive Waterfront Plan: Reclaiming the City's Edge* [1992], 23).

79. Thomas W. Sanchez et al., *Moving to Equity: Addressing Inequitable Effects of Transportation Policies on Minorities* (2003), www.civilrightsproject.harvard.edu (accessed Dec. 1, 2008), vii.

80. Robert R. M. Verchick, "In a Greener Voice: Feminist Theory and Environmental Justice," *Harvard Women's Law Journal* 19 (1996): 23, 65.

81. Catherine O'Neill, *Stanford Environmental Law Journal* 19 (2000): 3, 46-47.

82. James Flynn, Paul Slovic, and C. K. Mertz, "Gender, Race, and Perception of Environmental Health Risks," *Risk Analysis* 14 (1994): 1101, 1103.

83. California Office of Planning and Research, *General Plan Guidelines, Preliminary Draft* (2002), www.opr.ca.gov (accessed on Aug. 12, 2008), 20-21.

84. Executive Order 12, 898, 59 Fed. Reg. 7,629 (1994).

85. Ibid.

86. U.S. Government Accountability Office, "EPA Should Devote More Attention to Environmental Justice When Developing Clean Air Rules," GAO-05-289 (July 2005), www.gao.gov (accessed Dec. 1, 2008).

87. Office of the Inspector General, Environmental Protection Agency, "EPA Needs to Conduct Environmental Justice Reviews of Its Programs, Policies, and Activities," report no. 2006-P-00034 (Sept. 18, 2006).

88. Sur Contra La Contaminacion v. EPA, 202 F.3d 443, 449-450 (1st Cir. 2000); Morongo Band of Mission Indians v. FAA, 161 F.3d 569, 575 (9th Cir. 1998).

89. See Rebecca M. Bratspies, et al., *By the Stroke of a Pen: Seven Crucial Protections for Public Health and the Environment the New President Can Accomplish in the*

First 100 Days of His Administration (2008), Center for Progressive Reform www
.progressivereform.org (accessed Dec. 16, 2008), 18–22.

90. See, e.g., R.I.S.E. v. Kay, 768 F. Supp. 1144 (E.D. Va. 1991); East Bibb Twiggs
Neighborhood Association v. Macon-Bibb County Planning and Zoning Commission, 706 F. Supp. 880 (M.D. GA. 1989); Bean v. Southwestern Waste Management Corporation, 482 F. Supp. 673 (S.D. Texas 1979), affirmed without opinion,
780 F.2d 1038 (5th Cir. 1986).

91. Title VI, sec. 2000d.

92. Alexander v. Sandoval, 532 U.S. 275, 281 (2001).

93. Robert D. Bullard, "Environmental Justice in the 21st Century," in *The Quest for
Environmental Justice: Human Rights and the Politics of Pollution*, ed. Robert D.
Bullard (San Francisco: Sierra Club Books, 2005), 19, 36 (table 1.1).

94. Alexander v. Sandoval, 532 U.S. 275 (2001); see generally Bradford C. Mank, "Is
There a Private Cause of Action under EPA's Title VI Regulations? The Need to
Empower Environmental Justice Plaintiffs," *Columbia Journal of Environmental
Law* 24 (1999): 1.

95. See South Camden Citizens in Action v. New Jersey Dept. of Env. Protection, 274
F.3d 771 (3d Cir. 2001). See also Save Our Valley v. Sound Transit, 335 F.3d 932
(9th Cir. 2003).

96. Barry Commoner, *The Closing Circle: Nature, Man and Technology* (New York:
Knopf, 1971), 39.

6. Disaster Justice

1. Manual Pastor et al., *In the Wake of the Storm: Environment, Disaster, and Race
after Katrina* (2006), www.russellsage.org (accessed May 28, 2008), 9.

2. See Rena Steinzor and Christopher Schroeder, eds., *A New Progressive Agenda
for Public Health and the Environment* (Durham, N.C.: Carolina Press, 2004),
102–106.

3. Ibid.

4. See generally, Robert R. M. Verchick, "In a Greener Voice: Feminist Theory and
Environmental Justice," *Harvard Women's Law Journal* 19 (1996): 23.

5. U.S. Census Bureau, *Percent of Population, 2000—One Race: Black or African
American*, www.census.gov (accessed June 14, 2008). On average, the Atlantic hurricane season produces 10, of which about 6 become hurricanes and 2–3 become
major hurricanes. Manav Tanneeru, "It's Official: 2005 Hurricanes Blew Records
Away," *CNN.com*, Dec. 30, 2005, www.cnn.com (accessed June 14, 2008).

6. Before the storm, African Americans represented 32 percent of the population of
Louisiana, 36 percent of the population of Mississippi, and 26 percent of the population of Alabama. Pastor et al., *In the Wake of the Storm*, 3. Since the storm, those
numbers have not changed significantly.

7. Ibid.,9.

8. Ibid.

9. See Alyson Flournoy et al., *An Unnatural Disaster: The Aftermath of Hurricane Katrina* (2005), Center for Progressive Reform, (showing map of flooded areas in greater New Orleans). www.progressivereform.org (accessed Dec. 1, 2008), 36.

10. See, generally, Debra Lyn Bassett, "The Overlooked Significance of Place in Law and Policy: Lessons from Hurricane Katrina," 49, in Robert D. Bullard and Beverly Wright, eds., *Race, Place and Environmental Justice after Hurricane Katrina: Struggles to Reclaim, Rebuild, and Revitalize New Orleans and the Gulf Coast* (Boulder, Colo.: Westview Press, 2009); Richard Campanella, *Bienville's Dilemma: A Historical Geography of New Orleans* (Lafayette: University of Louisiana at Lafayette Press, 2008).

11. Reilly Morse, *Environmental Justice through the Eye of Katrina* (2008), Joint Center for Political and Economic Studies Health Policy Institute, www.jointcenter .org (accessed Dec. 1, 2008): 6.

12. Ibid.

13. Matthew Brown, "Just Close MR-GO, Corps Is Urged," *New Orleans Times-Picayune*, Sept. 5, 2006, A1.

14. Richard Campanella, *Geographies of New Orleans: Urban Fabrics before the Storm* (Lafayette: University of Louisiana at Lafayette Press, 2006), 61–62.

15. In re Katrina Canal Breaches Consolidated Litigation (E.D. La., No. 05-4182, Nov. 18, 2009).

16. Kaid Benfield et al., *After Katrina: New Solutions for Safe Communities and a Secure Energy Future* (New York: Natural Resources Defense Council, 2005).

17. Turner v. Murphy Oil USA, Inc., 472 F.Supp. 2d 830, 835 (E.D. La. 2007).

18. Steve Ritea et al., "Some Parts of Orleans Could Open on Monday," *New Orleans Times-Picayune*, Sept. 14, 2005, A1.

19. Altamont Environmental, Inc., *Sediment and Surface Water Sampling and Analyses*, Oct. 6, 2005 (reporting on soil and water samples taken from a location in the Bywater neighborhood of New Orleans; a location near the Industrial Canal in New Orleans; a location in Chalmette, Louisiana; and a location in Meraux, Louisiana) (on file with the author); see, generally, Rachel Godsil, Albert Huang, and Gina Solomon, "Contaminants in the Air and Soil in New Orleans after the Flood: Opportunities and Limitations for Community Empowerment," in Bullard and Wright, *Race, Place and Environmental Justice*, 115.

20. Aimee Cunningham, "Disaster's Consequences: Hurricane's Legacy Includes Arsenic," *Science News* 171, no. 5 (February 2005): 67.

21. Brajesh Dubey et al., "Quantities of Arsenic-treated Wood in Demolition Debris Generated by Hurricane Katrina," *Environmental Science and Technology* 41, no. 5 (Jan. 24, 2007): 1533–1536.

22. Morse, "Environmental Justice," 15.

23. Piers Blaikie et al., *At Risk: Natural Hazards, People's Vulnerability, and Disasters* (New York: Routledge, 1994), 9.

24. Susan L. Cutter, *The Geography of Social Vulnerability: Race, Class, and Catastrophe* (Sept. 23, 2005), understandingkatrina.ssrc.org (accessed Sept. 1, 2008).

25. Flournoy, *Unnatural Disaster*, 35.

26. Ibid.

27. Ibid.

28. Ibid.

29. Ibid.

30. Bill Walden, "Katrina Scatters a Grim Diaspora," *BBC News*, Sept. 1, 2005, www.bbc.co.uk (accessed July 10, 2008).

31. Pastor et al., *In the Wake of the Storm*, 22 (citing studies).

32. Ibid.

33. Ibid.

34. My account of the Gretna bridge standoff draws from Kristin E. Henkel et al., "Discrimination, Individual Racism, and Hurricane Katrina," *Analysis of Social Issues and Public Policy* 6 (2006): 99; "The Bridge to Gretna," Dec. 18, 2005, *60 Minutes*, CBS News, transcript, www.cbsnews.com (accessed Feb. 26, 2008); "Residents Returned to New Orleans Told to Evacuate," *American Morning*, CNN, Sept. 20, 2005, transcript, transcripts.cnn.com (accessed Jan. 19, 2010). Andrew Buncombe, "Racist Police Blocked Bridge and Forced Evacuees Back at Gunpoint," *Independent*, Sept. 11, 2005, www.independent.co.uk (accessed Sept. 22, 2008); Gardner Harris, "Police in Suburbs Blocked Evacuees, Witnesses Report," *New York Times*, Sept. 10, 2005, A13; Clip Johnson, "Police Made Their Storm Misery Worse," *San Francisco Chronicle*, Sept. 9, 2005, B1; Shaun Waterman, "Cops Trapped Survivors in New Orleans," UPI, Sept. 9, 2005, www.washtimes.com (accessed Jan. 11, 2009). Larry Bradshaw and Lorrie Beth Slonsky, "Hurricane Katrina—Our Experiences," Sept. 5, 2005, sfsocialists.livejournal.com (accessed Sept. 22, 2008); Chris Kirkham, "Bridge Blockade after Katrina Remains a Divisive Issue," *New Orleans Times-Picayune* Sept. 2, 2007, National sec.; "Gretna, Louisiana," (2005) City-data.com, www.city-data.com (accessed Sept. 1, 2008).

35. Buncombe, "Racist Police Blocked Bridge" (quoting Larry Bradshaw).

36. Ibid. (quoting Arthur Lawson, chief of police, Gretna, Louisiana).

37. PolicyLink, *A Long Way Home: The State of Housing Recovery in Louisiana* (2008), www.policylink.org (accessed Dec. 1, 2008), 6.

38. Ibid., 42.

39. Ibid., 47; See, generally, Lisa K. Bates and Rebekah A. Green, "Housing Recovery in the Ninth Ward: Disparities in Policy, Process, and Prospects," in Bullard and Wright, *Race, Place and Environmental Justice*, 229.

40. Erica Williams et al., *The Women of New Orleans and the Gulf Coast: Multiple*

Disadvantages and Key Assets for Recovery, pt. 1 (Aug. 2006), Institute for Women's Policy Research, www.iwpr.org (accessed May 28, 2008), 5 (using 2004 figures).

41. Sarah Vaill, *A Calm in the Storm: Women Leaders in Gulf Coast Recovery* (2006), Women's Funding Network and Ms. Foundation for Women, www.ms.foundation.org (accessed May 28, 2008), 3.

42. See ibid. (citing statistics).

43. Before Katrina, African Americans made up about 67 percent of the population of Orleans Parish. U.S. Census Bureau, *State and County Quick Facts: Orleans Parish, Louisiana* (undated, quickfacts.census.gov (accessed May 28, 2008) (using 2005 pre-Katrina data). The phrase "chocolate city" was coined by funk legend George Clinton's 1975 classic of the same name, which he recorded with his band Parliament. See Parliament, "Chocolate City," *Chocolate City* (1975), Island Def Jam Music Group, 2003.

44. Michelle Krupa, "Black, Hispanic Numbers Jump," *New Orleans Times-Picayune*, May 14, 2009, A1.

45. Vaill, "Calm in the Storm," 3.

46. Ibid.

47. Ibid.

48. Children's Health Fund, *Legacy of Shame: The On-going Public Health Disaster of Children Struggling in Post-Katrina Louisiana* (Nov. 4, 2008), Mailman School of Public Health, Columbia University, www.childrenshealthfund.org (accessed Dec. 1, 2008).

49. Peter Ortiz, "Deportation Fears Keep Latinos from Seeking Katrina Aid," *DiversityInc.com*, Sept. 9, 2005, www.diversityinc.com (accessed Dec. 1, 2008).

50. Pastor et al., "Wake of the Storm," 22.

51. Ibid.

52. Mark Waller, "In the Wake of Katrina, Thousands of Spanish-speaking People Are Migrating to New Orleans, Drawn by the Dream of a Better Life," *New Orleans Times-Picayune*, Oct. 8, 2006, National sec.

53. Krupa, "Black, Hispanic Numbers Jump" (reporting that "[w]hile [Louisiana's] population as a whole inched up just 0.9 percent between 2007 and 2008, growth among Hispanics leapt 4.8 percent").

54. *See, generally*, Laurel E. Fletcher, et al., Rebuilding after Katrina: A Population-Based Study of Labor and Human Rights in New Orleans, June 2006 (project of research centers at University of California-Berkeley and Tulane University), *available at* www.law.berkeley.edu; Judith Browne-Dianis, et al., And Injustice for ALL: Workers' Lives in the Reconstruction of New Orleans (undated) (project of Advancement Project), *available at* www.advancementproject.org.

55. C. Stone Brown, "Katrina's Forgotten Victims: Native American Tribes," Pacific News Service, Sept. 11, 2005, www.pacificnews.org (accessed July 18, 2008);

Michael Eric Dyson, *Come Hell or High Water: Hurricane Katrina and the Color of Disaster* (New York: Basic Civitas, 2006), 143.

56. "Isolated Native American Communities Struggle in the Aftermath of Hurricanes Katrina and Rita," interview by Gregory Berger with Marlene Foret, Grand Caillou/Dulac Band of the Biloxi-Chitimachas; Albert Naquin, Isle de Jean Charles Band of the Biloxi-Chitimachas; and Randy Verdun, Isle de Jean Charles Band of Biloxi-Chitimachas, *Democracy Now*, Nov. 24, 2005, www.democracynow .org (accessed July 18, 2008).

57. Ibid.

58. Dyson, *Come Hell or High Water*, 143–144.

59. Matt Stiles and Chase Davis, "Ike Damaged More Than Half of Houston's Apartments," *Houston Chronicle*, Oct. 21, 2008, www.chron.com (accessed Jan. 11, 2009).

60. Rhiannon Myers, "Public Housing Residents Angry, Confused," *Galveston Daily News*, Sept. 24, 2008, www.galvestondailynews.com (accessed Jan. 14, 2008).

61. Office of Response and Restoration, National Oceanic and Atmospheric Administration, "Hurricane Ike Response," press release, Sept. 16, 2008 (last revised Sept. 26, 2008), www.response.restoration.noaa.gov (accessed Dec. 17, 2008).

62. Dane Schiller et al., "Needy Residents Desperate for Relief," *Houston Chronicle*, Sept. 16, 2008, A1.

63. Jennifer Peltz, "Hurricane Barriers Floated to Keep Sea out of NYC," *Washington Post*, May 30, 2009, www.washingtonpost.com (accessed June 5, 2009). In fact, scientists and engineers have been developing barrier plans to protect New York City and parts of New Jersey from serious storms. The most ambitious project would involve a 5-mile-long barrier between New Jersey and Queens. Ibid.

64. Jared Bernstein et al., "U.S. Poverty Policy in the Aftermath of Hurricane Katrina" (2006), Russell Sage Foundation, www.russellsage.org (accessed Dec. 28, 2008), 9, table 1. The ratings are based on statistics from the year 2000.

65. Compare A. C. Thompson, "Katrina's Hidden Race War," *Nation*, Dec. 17, 2008, www.thenation.com (accessed Dec. 18, 2008) (reporting that after Katrina, white vigilante groups patrolled New Orleans, blockaded streets, and shot at least 11 black men).

66. United States Census Bureau, "Table B08201 Household Size by Vehicles Available- Universe Households," in *American Community Survey* (2004), factfinder .census.gov (accessed Jan. 21, 2010).

67. Mary Wollstonecraft, *A Vindication of the Rights of Women*, ed. Carol H. Poston (New York: Norton, 1975), 71.

68. Jonathan Tilove, "Author Paints Hurricane as Clarion Call on Poverty" (interview with Michael Eric Dyson), *New Orleans Times-Picayune* Feb. 12, 2006, National sec.

69. See, e.g., Jean-Jacque Rousseau, *The Social Contract and the First and Second*

Discourses, ed. and trans. Susan Dunn (New Haven, Conn.: Yale University Press, 2002) (describing social contract based on popular sovereignty); John Locke, *Locke: Two Treatises on Government*, ed. Peter Laslett (Cambridge: Cambridge University Press, 1988) (describing social contract based on protections of life, liberty, and property); John Rawls, *A Theory of Justice* (1971) (Oxford: Oxford University Press, 1973) (describing social contract based on Kantian precepts of fairness).

70. Michael Ignatieff, "The Broken Contract," *New York Times Magazine*, Sept. 25, 2005, 15.

71. Ibid.

72. Indeed, in January 2005, parties to the so-called Hyogo Framework for Action (see Chapter 7) agreed to the following language: "States have the primary responsibility to protect the people and property on their territory from hazards and . . . to give high priority to disaster risk reduction in national policy, consistent with their capacities and resources available to them." *Hyogo Framework for Action 2005–2015: Building the Resilience of Nations and Communities to Disaster* (2005) www.unisdr.org (accessed Dec. 1, 2008).

73. See Flournoy "Unnatural Disaster."

74. Pastor et al., "Wake of the Storm," 5–6.

75. Ibid.

76. Ronald Reagan, Inaugural Address (Jan. 20, 1981), www.reaganlibrary.com (accessed July 20, 2008).

77. See, generally, David Harvey, *A Brief History of Neoliberalism* (Oxford: Oxford University, 2005).

78. Ibid., 2.

79. For instance, in the closing days of the 2008 presidential election, Barack Obama made the following case: "Now, understand, I don't believe that government can or should try to solve all our problems. . . . But I do believe that government should do that which we cannot do for ourselves—protect us from harm; provide a decent education for all children—invest in new roads and new bridges, in new science and technology." "Remarks of Senator Barack Obama—As Prepared for Delivery, Canton, Ohio, Oct. 27, 2008" (known as the Closing Argument Speech), thepage.time.com (maintained by Time in partnership with CNN) (accessed Jan. 9, 2009).

80. Dag Einar Thorson and Amund Lie, "What Is Neoliberalism?" 2006 (describing neoliberalism with reference to the work of Milton Friedman, Robert Nozick, and Friedrich Hayek, among others), unpublished manuscript on file with the author.

81. See Robert Nozick, *Anarchy, State, and Utopia* (New York: Basic Books, 1974), 172–73.

82. Stephen Graham, "Cities under Siege: Katrina and the Politics of Metropolitan America," Sept. 12, 2005, understandingkatrina.ssrc.org (accessed Dec. 28, 2008).

83. Ibid.

84. President George W. Bush, "Remarks at National Day of Prayer and Remembrance Service at Washington National Cathedral, Washington, D.C.," Sept. 16, 2005, www.whitehouse.gov (accessed July 20, 2008).

85. Bernstein et al., "Poverty Policy."

86. John Tierney, "Ben Franklin Had the Right Idea for New Orleans," opinion, *New York Times*, Sept. 3, 2005, A21.

87. Michelle Krupa, "We've Been There," *New Orleans Times-Picayune* July 20, 2008, National sec.

88. Ibid.

89. Rockefeller Foundation/Time, *Campaign for American Workers Survey, Executive Summary* (July 17, 2008), www.rockfound.org (accessed July 28, 2008), 5.

90. Naomi Klein, *The Shock Doctrine: The Rise of Disaster Capitalism* (New York: Metropolitan Books, 2007).

91. Ibid., 387.

92. Ibid., 398.

93. ActionAid International, *Tsunami Response: A Human Rights Assessment* (Jan. 2006), www.actionaid.org (Jan. 12, 2009), 9.

94. Ibid., 410.

95. Ibid., 410; see also "Pro-free-market Ideas for Responding to Hurricane Katrina and High Gas Prices," in e-mail, Rep. Paul Teller to multiple recicpients Sept. 13, 2005, (later obtained and made public by Naomi Klein), www.naomiklein.org (accessed September 20, 2008).

96. Charles Babington, "Some GOP Legislators Hit Jarring Notes in Addressing Katrina," *Washington Post*, Sept. 10, 2005, A4 (emphasis added); see also Mtanguliza Sanyika, "Katrina and the Condition of Black New Orleans: The Struggle for Justice, Equity, and Democracy," in Bullard and Wright, *Race, Place and Environmental Justice*, 87, 94–95, (referring to policies that discouraged poor people from returning as "Katrina cleansing"). Christopher Cooper, "Old-line Families Escape Worst of Flood and Plot the Future," *Wall Street Journal*, Sept. 8, 2005, A1.

97. About a week after the flood, the *Wall Street Journal* reported on the efforts of a few old-line families to take advantage of what they saw as the city's new blank slate. The piece devoted particular attention to a lawyer named James Reiss, a flamboyant character who, days after the flood, had "helicoptered in an Israeli security company to guard his Audubon house and those of his neighbors." According to the article, "[t]he power elite of New Orleans—whether they are still in the city or have moved temporarily to enclaves such as Destin, Fla., and Vail, Colo.—insist the remade city won't simply restore the old order. New Orleans before the flood was burdened by a teeming underclass, substandard schools and a high crime rate. The city has few corporate headquarters. The new city must be something different, Reiss says, with better services and fewer poor people." Cooper, "Old-line Families Escape Worst of Flood."

98. Elisabeth Bumiller and Anne E. Kornblut, "Storm and Crisis: Political Memo; Black Leaders Say Storm Forced Bush to Confront Issues of Race and Poverty," *New York Times*, Sept. 17, 2005, sec. 1.

99. President George W. Bush and Lt. General Russell L. Honoré, press conference, Sept. 12, 2005, www.whitehouse.gov (accessed on Nov. 12, 2008).

100. President George W. Bush, Remarks at National Day of Prayer and Remembrance Service at Washington National Cathedral, Washington, D.C., Sept. 16, 2005, transcript, www.whitehouse.gov (accessed Oct. 1, 2008).

101. See Mary Elise DeCoursey, "St. Bernard Parish on the Hook for Legal Fees in Rental Ordinance Fight," *New Orleans Times-Picayune*, July 3, 2008, www.nola .com (accessed Jan. 11, 2008).

102. John Pope, "Evoking King, Nagin Calls N.O. 'Chocolate' City," *New Orleans Times-Picayune*, Jan. 17, 2006, National sec. (quoting Ray Nagin).

103. See, generally, Nancy Levit and Robert R. M. Verchick, *Feminist Legal Theory: A Primer* (New York: New York University Press, 2006).

104. Richard Thompson Ford, *The Race Card: How Bluffing about Bias Makes Race Relations Worse* (New York: Farrar, Straus and Giroux, 2008), 37.

105. Ibid., 57.

106. Lisa de Moraes, "Kanye West's Torrent of Criticism, Live and on NBC," *Washington Post*, Sept. 3, 2005, C1 (recounting the episode).

107. Ford, *Race Card*, 41 ("Kanye West called the president of the United States a racist").

108. Jacob Weisberg, "An Imperfect Storm: How Race Shaped Bush's Response to Katrina," *Slate*, Sept. 7, 2005, www.slate.com (accessed Oct. 3, 2008).

109. In the 2000 election, Bush won 9 percent of the black vote. Elisabeth Bumiller and Anne E. Kornblut, "Black Leaders."

110. Ibid.

111. See, generally, Derrick A. Bell, Jr., "Brown v. Board of Education and the Interest-convergence Dilemma," *Harvard Law Review* 93 (1980): 518.

112. My description of this study draws from Shanto Iyengar and Kyu S. Hahn, "Natural Disasters in Black and White: How Racial Cues Influenced Public Response to Hurricane Katrina," June 10, 2007, unpublished manuscript on file with the author, and Shanto Iyengar and Richard Morin, "Natural Disasters in Black and White," *Washington Post*, June 8, 2006, www.washingtonpost.com (accessed Jan. 12, 2009).

113. See, for instance, Charles R. Lawrence, III, "The Id, the Ego, and Equal Protection: Reckoning with Unconscious Racism," *Stanford Law Review* 39 (1987): 317.

114. Edward Patrick Boyle, "It's Not Easy Bein' Green: The Psychology of Racism, Environmental Discrimination, and the Argument for Modernizing Equal Protection Analysis," *Vanderbilt Law Review* 46 (1993): 937.

115. See, e.g., Jerry Kang, "Trojan Horses of Race," *Harvard Law Review* 118 (2005):

148 9. Linda Hamilton Krieger, "The Content of Our Categories: A Cognitive Bias Approach to Discrimination and Equal Employment Opportunity," *Stanford Law Review* 47 (2005): 1161.

116. Project Implicit, implicit.harvard.edu (accessed Dec. 17, 2008).

117. Project Implicit, *General Information*, implicit.harvard.edu (accessed Dec. 17, 2008).

118. Thierry Devos et al., "Is Barack Obama American Enough to Be the Next President? The Role of Ethnicity and National Identity in American Politics" (2008), unpublished, www.rohan.sdsu.edu (accessed Dec. 17, 2008).

119. Thierry Devos and Debbie S. Ma, "Is Kate Winslet More American Than Lucy Liu? The Impact of Construal Processes on the Implicit Ascription of a National Identity," *British Journal of Social Psychology* 47, no. pt. 2 (2008): 191–215. During the 2008 election, this study and the one involving Obama were described in Nicholas D. Kristof, "What? Me Biased?" *New York Times*, Oct. 29, 2008, A39.

120. Krieger, "Content of Our Categories," 1167.

121. Washington v. Davis, 426 U.S. 229 (1976).

122. Krieger, "Content of Our Categories," 1167.

123. See ibid.

124. Ibid., 1216–1217.

7. Winds of Change

1. Manual Pastor et al., *In the Wake of the Storm: Environment, Disaster, and Race after Katrina* (2006), www.russellsage.org (accessed May 28, 2008), 27.

2. Ibid., 33.

3. Richard Thompson Ford makes the same point with regard to helping those harmed by institutional racism: *The Race Card: How Bluffing about Bias Makes Race Relations Worse* (New York: Farrar, Straus and Giroux, 2008), 347–348.

4. Jared Bernstein et al., *U.S. Poverty Policy in the Aftermath of Hurricane Katrina*, (2006), www.russellsage.org (accessed Dec. 28, 2008), 3.

5. Rena Steinzor and Margaret McClune, The Hidden Lesson of the Vioxx Fiasco: Reviving a Hollow FDA (Oct. 2005), Center for Progressive Reform, www.progressivereform.org (accessed Dec. 1, 2008), 1.

6. President Barack Obama, interview by Matt Lauer, *Sunday Today*, Feb. 2, 2009, transcript, www.msnbc.com (accessed June 9, 2009); see also "FDA Widens Peanut Butter Salmonella Probe," *Forbes*, Jan. 15, 2009, www.forbes.com (accessed June 9, 2009); Josh Funk, "Peanut Butter Recalled over Salmonella," *Washington Post*, Feb. 15, 2007, www.washingtonpost.com (accessed June 9, 2009).

7. David Barboza, "As More Toys Are Recalled, Trail Ends in China," *New York Times*, June 19, 2007, www.nytimes.com (accessed June 9, 2009).

8. Alyson Flournoy et al., "An Unnatural Disaster: The Aftermath of Hurricane Katrina" (2005), Center for Progressive Reform, www.progressivereform.org (accessed Dec. 1, 2008), 4.

9. Shaila Dewan, "Tennessee Ash Flood Larger Than Initial Estimate," *New York Times*, Dec. 27, 2008, A10. "Environmentalists Fear Risks from Tennessee Ash Spill," *Washington Post*, Dec. 26, 2008, A20.

10. Toxics Use Reduction Institute, *TURA Data: A Community Guide to Toxics Information from Massachusetts' Toxics Use Reduction Act* (2008), www.turadata .turi.org (accessed Dec. 8, 2008).

11. See, e.g., U.S. Rep. Gene Green, "Green 'Deeply Concerned' about Ike's Effect on Local Superfund Site, Asks EPA to Investigate," press release, Sept. 24, 2008, www.house.gov (accessed Dec. 1, 2008).

12. Rena Steinzor and Margaret McClune, *The Toll of Superfund Neglect: Toxic Waste Dumps and Communities at Risk* (2007), Center for Progressive Reform and Center for American Progress, www.progressivereform.com (accessed Dec. 1, 2008).

13. See ibid., 2, fig. 1.

14. Ibid., 3.

15. HurricaneCity, "Pensacola, Florida's History with Tropical Storm Systems" www .hurricanecity.com (accessed Dec. 1, 2008) (drawing from newspaper and federal government resources).

16. GlynnCounty.com, *Hurricanes* (2008), www.glynn.county.com (accessed Dec. 8, 2008).

17. See, e.g., U.S. Geological Survey Earthquake Hazards Program, *Earthquake List for Map California Nevada* (updated December 8, 2008), www.earthquake.usgs .gov (accessed Dec. 8, 2008).

18. Green, "Green 'Deeply Concerned.'"

19. See 44 CFR 7.1–7.949.

20. 42 U.S.C. sec. 5151 (2008); see also Daniel A. Farber, "Disaster Law and Inequality," *Journal of Law and Inequality* 25 (2007): 297, 310–311 (discussing Section 308).

21. 42 U.S.C. sec. 5151. The categories of "disability" and "English proficiency" were added to Section 308 in the aftermath of Hurricane Katrina. See Post-Katrina Emergency Management Reform Act of 2006, P.L. 109–295, sec. 389a.

22. See Cannon v. University of Chicago, 441 U.S. 677, 677 (1979) (citing Cort v. Ash, 422 U.S. 66 [1975]) (holding that legislative intent and phrasing that identifies a benefited group may be used, along with other information, to infer private remedy).

23. 42 U.S.C. sec. 512(b)(4) (2008).

24. 42 U.S.C. sec. 512(b)(5) (2008).

25. See U.S. Government Accountability Office, *Transportation-disadvantaged Populations: Actions Needed to Clarify Responsibilities and Increase Preparedness for Evacuations*, GAO-07-44 (Dec. 2006), www.gao.gov (accessed Dec. 28, 2008).

26. Ibid.

27. This discussion draws on a policy paper I coauthored that recommended improvements to Clinton's original Executive Order on Environmental Justice. See

Rebecca M. Bratspies et al., *Protecting Public Health and the Environment by the Stroke of a Presidential Pen: Seven Executive Orders for the President's First One Hundred Days* (November 2008), Center for Progressive Reform, www.progressivereform.org (accessed on Dec. 1, 2008).

28. See, e.g., Van Johnson, *Green Collar Economy: How One Solution Can Fix Our Two Biggest Problems* (New York: Harper One, 2008); Bratspies et al., *Protecting Public Health*, 21.

29. ACORN v. U.S. Army Corps of Engineers, 2000 WL 433332 (E.D. La. 2000) affirmed per curiam, 245 F. 3d 790 (5th Cir. 2000).

30. Farber, "Disaster Law and Inequality," 315.

31. Ibid., 318–319.

32. See Dylan J. McDonald, *The Teton Dam Disaster* (Mount Pleasant, S.C.: Arcadia, 2006); Pierce O'Donnell, "Leave No Katrina Victims Behind," *Huffington Post*, Aug. 29, 2008, www.huffingtonpost.com (accessed Dec. 1, 2008); Center for Land Use Interpretation, *Teton Dam Failure Site*, ludb.clui.org (accessed Dec. 1, 2008).

33. U.S. General Accounting Office, *FEMA Cerro Grande Claims: Payments Properly Processed, but Reporting Could Be Improved*, GAO-04-129 (Dec. 24, 2003), www.gao.gov (accessed Dec. 1, 2008).

34. Ibid.; National Park Service, *Cerro Grande Prescribed Fire: Board of Inquiry Final Report* (Feb. 6, 2001), www.nps.gov (accessed Dec. 1, 2008); U.S. Federal Emergency Management Agency, "FEMA Begins Work on New Cerro Grande Fire Compensation," press release, July 17, 2000, www.fema.gov (accessed Dec. 1, 2008).

35. Farber, "Disaster and Inequality," 318–319.

36. Ibid., 319.

37. I recognize, here, that I am treading a thin line. While I do not believe that personal characteristics like skin color or ethnicity should *constrain* benefits (whether for reasons of conscious intent or unconscious bias), I do think such characteristics should play a conscious role in decisions to *augment* benefits (like bus service or multilanguage announcements) for people for whom such benefits would be useful. That is, I believe it is *sometimes* appropriate to use factors like race, ethnicity, sex, or income—alone or combined—as a proxy for social vulnerability; and it is appropriate to use social vulnerability as a reason for *augmenting* the range of disaster services and aid provided. Skeptics will claim that increasing a resource for some groups necessarily means decreasing it for other groups, and that I am therefore condoning the use of skin color and ethnicity to decrease resources available to whites and European Americans. I have two responses. First, as long as we are talking about industrialized countries like the United States, it seems unlikely that we could not provide for the needs of both the advantaged and disadvantaged. Second, even if I am wrong about that, I believe justice requires helping vulnerable people more. Insofar as this results in a disproportional division of resources based on skin color or ethnicity, it can be justified in the same way that traditional affirmative action programs have been justified. Some readers, of course, may

disapprove of affirmative action for a variety of intelligent and well-intended reasons. On that point, we must politely agree to disagree.

38. International Covenant on Civil and Political Rights, 999 U.N.T.S. 171 (1976); Protocol Relating to the Status of Refugees, 606 U.N.T.S. 267 (1967).

39. International Covenant on Economic, Social and Cultural Rights 993 U.N.T.S. 3 (1966), Convention on the Elimination of All Forms of Discrimination against Women 1249 U.N.T.S. 13 (1979); Convention on the Rights of the Child, 1577 U.N.T.S. 3 (1989).

40. See generally, *Hyogo Framework for Action 2005–2015: Building the Resilience of Nations and Communities to Disaster* (2005), www.unisdr.org (accessed Dec. 1, 2008).

41. United Nations Secretariat of the International Strategy for Disaster Reduction (UN/ISDR), *Indicators of Progress: Guidance on Measuring the Reduction of Disaster Risks and the Implementation of the Hyogo Framework for Action* (Jan. 2008), www.unisdr.org (accessed Dec. 1, 2008).

42. Ibid.

43. The facts and examples offered here are drawn from U.N. International Strategy for Disaster Reduction, *Implementation of the International Strategy for Disaster, Risk Reduction* (Sept. 5, 2007), Report of the Secretary General, www.unisdr.org (accessed Dec. 1, 2008); U.N. International Strategy for Disaster Reduction, *Linking Disaster Risk Reduction and Poverty Reduction: Good Practices and Lessons Learned* (2008), www.unisdr.org (accessed Dec. 1, 2008); and UN/ISDR, *Hyogo Framework for Action 2005–2015: Building the Resilience of Nations and Communities to Disasters* (undated), www.unisdr.org (accessed Dec. 1, 2008).

44. UN/ISDR, "Linking Disaster Risk Reduction," 1.

45. Oxfam International, *Rethinking Disasters: Why Death and Destruction Is Not Nature's Fault but Human Failure* (2008), www.oxfam.org (accessed May 28, 2008).

46. U.N. Office of the High Commissioner for Human Rights, *Guiding Principles on Internal Displacement* (undated) www.unhchr.ch (accessed Dec. 1, 2008). The U.N. General Assembly has formally recognized this document at least twice. In a 2003 resolution, the Assembly noted "that the protection of internally displaced persons has been strengthened by identifying, reaffirming and consolidating specific standards for their protection, in particular through the Guiding Principles on Internal Displacement." UN Doc. A/RES/58/177, preambular paragraphs. In 2005 the Assembly approved the World Summit Outcome document, which declared, "We recognize the Guiding Principles on Internal Displacement as an important international framework for the protection of internally displaced persons." Draft Outcome Document (2005), www.un.org (accessed Dec. 1, 2008), para. 132. See Frederic L. Kirgis, "Hurricane Katrina and Internally Displaced Person," *ASIL Insights*, (Sept. 21, 2005), American Society of International Law, www.asil.org (accessed Dec. 1, 2008).

47. Francis M. Deng, *Introductory Note by the Representative of the Secretary-General on Internally Displaced Persons* (undated), U.N. Office for the Coordination of Humanitarian, www.reliefweb.int (accessed Dec. 28, 2008).

48. U.N. Guiding Principles on Internal Displacement U.N. Doc. E/CN. 4/1998/53/ Add.2 (1998), principle 1. Discussion and quotations of the principles are drawn from principles 1–30.

49. Bahame Tom Nyanduga, "The Challenge of Internal Displacement in Africa," *Forced Migration Review* 21 (September 2004): 58.

50. This song, popular among folksingers, "is commonly mistaken as being of Irish origin"; liner notes, Déanta, *Ready for the Storm*, Green Linnet, 1994. "The Lakes of Pontchartrain" was likely introduced to Europe by British and French soldiers returning from Louisiana after the War of 1812. See liner notes, Planxty, *Cold Blow the Rainy Night* (1974), Sanachie, 1989. For more information on this song, see Barry Taylor, *The Lakes of Pontchartrain* (undated) www.geocities.com (accessed July 7, 2008). A video of Bob Dylan performing the song in 1989 is available on YouTube (accessed July 7, 2008).

51. "The Lakes of Pontchartrain," in *Traveling for Love: Song Book* (Queensland: State Library of Queensland, 2008), www.slq.qld.gov.au (accessed Dec. 5, 2008).

8. Precaution and Social Welfare

Epigraph. Alfred Korzybski, *Science and Sanity: An Introduction to Non-Aristotelian Systems and General Semantics* (Fort Worth, Tex.: Institute of General Semantics, 1994), 58.

1. Jorge Louis Borges, "On Exactitude in Science," in *Collected Fictions*, trans. Andrew Hurley (New York: Viking, 1998), 325.

2. Alfred Korzybski is perhaps best known for founding the discipline of general semantics, which seeks to improve decision-making through a more critical use of words and symbols. See, generally, Korzybski, *Science and Sanity*.

3. For several such maps, see *Hurricane Katrina Maps*, Perry Castañada Library Map Collection, University of Texas, www.lib.utexas.edu (accessed Jan. 10, 2009).

4. See, e.g., Edward R. Tufte, *Envisioning Information* (Los Angeles: Graphic Press, 1990). Edward R. Tufte, *Visual Explanations: Images and Quantities, Evidence and Narrative* (Los Angeles: Graphic Press, 1997). Once opened, these books are hard to put down.

5. For an image of the Underground commuter map, with commentary by Edward Tufte and others, see www.edwardtufte.com (accessed Dec. 18, 2008).

6. See Edward R. Tufte, *PowerPoint Does Rocket Science—and Better Techniques for Technical Reports* (Sept. 6, 2005), www.edwardtufte.com (accessed Dec. 18, 2008).

7. Richard A. Posner, *Catastrophe: Risk and Response* (New York: Oxford University Press, 2004), 21.

8. Lee Clarke, *Worst Cases: Terror and Catastrophe in the Popular Imagination* (Chicago: University of Chicago Press, 2006), xi.

9. Protection of the Environment Administration Act, New South Wales Consolidated Acts (1991), W.S.I. sec. 6(2)(a).

10. United Nations Framework Convention on Climate Change, *Climate Change Information Sheet* 19 (2000), www.unfccc.int (accessed Jan. 14, 2009).

11. Sidney A. Shapiro and Christopher H. Schroeder, "Beyond Cost-benefit Analysis: A Pragmatic Reorientation," *Harvard Environmental Law Review* 32 (2008): 433, 476.

12. Wendy E. Wagner, "The Triumph of Technology-based Standards," *University of Illinois Law Review* 2000 (2000): 83, 92; see also Shapiro and Schroeder, "Beyond Cost-benefit Analysis," 479; David M. Driesen, *The Feasibility Principle* (2004), Center for Progressive Reform, www.progressivereform.org (accessed Dec. 26, 2008), 1.

13. Driesen, "Feasibility Principle," 3–4.

14. See Douglas A. Kysar, "It Might Have Been: Risk, Precaution, and Opportunity Costs," *Journal of Land Use and Environmental Law* 22 (2006): 24 ("much of U.S. environmental law and regulation continues to be based instead on policies and procedures that reflect a precautionary approach").

15. 7 U.S.C. sec. 136a(d)(1)(C), (hh) (2000); see also Shapiro and Schroeder, "Beyond Cost-benefit Analysis," 480 (using this example).

16. Shapiro and Schroeder, "Beyond Cost-benefit Analysis," 481.

17. Adam B. Jaffe et al., "Environmental Regulation and the Competitiveness of U.S. Manufacturing: What Does the Evidence Tell Us?" in *Trade and the Environment: Economic, Legal, and Policy Perspectives*, ed. Alan M. Rugman, et al. (Northampton, Mass.: Edward Elgar, 1998), 132, 157 (reviewing economic data since 1970 and concluding that "[o]verall, there is little evidence to support the hypothesis that environmental regulations have had a large adverse effect on competitiveness, however that elusive term is defined"); see also Frank Ackerman and Lisa Heinzerling, *Priceless: On Knowing the Price of Everything and the Value of Nothing* (New York: New Press, 2004), 10 ("virtually no job losses can be traced to environmental regulation"); International Chemical Secretariat, *Cry Wolf: Predicted Costs by Industry in the Face of New Regulation* (April 2004), www.chemsec .org (accessed Jan. 10, 2009).

18. For discussions of how cultural values are linked to risk perception, see, generally, Dan M. Kahan, "Two Conceptions of Emotion in Risk Regulation," *University of Pennsylvania Law Review* 156 (2008): 741; Dan M. Kahan et al., "Fear of Democracy: A Cultural Evaluation of Sunstein on Risk," *Harvard Law Review* 119 (2006): 1071 (reviewing Cass R. Sunstein, *Laws of Fear: Beyond the Precautionary Principle*

[Cambridge: Cambridge University Press, 2005]); and James Flynn, Paul Slovic, and C. K. Mertz, "Gender, Race, and Perception of Environmental Health Risks," *Risk Analysis* 14 (1994): 1101.

19. Ackerman and Heinzerling, *Priceless*, 37.

20. When a worker is willing to give up a risky job for a lower-paying "safe" job, economists see that as evidence that the worker is willing to "pay" for increased safety in the workplace.

21. Ackerman and Heinzerling, *Priceless*, 61, 75–76.

22. See Lisa Heinzerling, "The Rights of Statistical People," *Harvard Environmental Law Review* 24 (2000): 189, 203 (describing a statistical life as an "aggregation of relatively small risks of harm to the individuals in a population"). The term "statistical life" may also be used in a slightly different way to describe "a life expected to be lost as a function of probabilities of death applied to a population of persons," where "the person expected to die is not identifiable in advance of death, and the probabilistic estimates are uncertain"; 196. The difference between these uses is not important to my analysis.

23. Shapiro and Schroeder, "Beyond Cost-benefit Analysis," 439; Frank Fischer, *Reframing Public Policy: Discursive Politics and Deliberative Practices* (New York: Oxford University Press, 2003), 118.

24. Shapiro and Schroeder, "Beyond Cost-benefit Analysis," 439.

25. 33 U.S.C. sec. 701(a). The provision went on to require that "the lives and social security of people" must also be at risk; ibid.

26. See, generally, U.S. Army Corps of Engineers, *Economic and Environmental Principles and Guidelines for Water and Related Land Resources Implementation Studies* (Mar. 10, 1983), www.usace.army.mil, 19–102, setting forth procedures for determining "net economic development," the cost-benefit measure used by the Corps); Donald Hornstein, et al., *Broken Levees: Why They Failed* (2005), Center for Progressive Reform, www.progressivereform.com (accessed May 13, 2009), 9 (noting that Army Corps guidelines direct analysts to "address the issue of prevention of loss of life when evaluating alternative plans, but they are not required to formally estimate the number of lives saved or lost").

27. Daniel A. Farber, *Eco-pragmatism: Making Sensible Environmental Decisions in an Uncertain World* (Chicago: University of Chicago Press, 1999), 120 (attributing use of modern cost-benefit analysis in public regulation to Robert McNamara and colleagues at Ford Motor Company). In addition to Robert McNamara, the Whiz Kids included Charles B. "Tex" Thorton, Wilbur Andreson, Charles Bosworth, J. Edward Lundy, Arjay Miller, Ben Mills, George Moore, Francis "Jack" Reith, and James Miller. For more on their influence over business and military thinking, see John Byrne, *The Whiz Kids: Ten Founding Fathers of American Business—and the Legacy They Left Us* (New York: Doubleday, 1993).

28. Executive Order No. 12,291, 46 Fed. Reg. 13,193 (Feb. 19, 1981).

29. Elena Kagan, "Presidential Administration," *Harvard Law Review* 114 (2001): 2245, 2281.

30. Executive Order No. 12,866, 58 Fed. Reg. 51,735 (Oct. 4, 1993).

31. The innovation in optimality was the switch from the Pareto principle to the Kaldor-Hicks efficiency test. Under the Pareto principle, a project or regulation is desirable if it makes at least one person better off and no person worse off. This almost never happens in real life; in addition, compensating losers is often believed to be too difficult and too expensive. The Kaldor-Hicks test, in contrast, approves of any project or regulation as long as the winners are enriched sufficiently that they *could* compensate the losers if they wanted to, *even though* no compensation is really required. This outcome is acceptable in social welfare terms, since it is the *sum* of individual utility rather than the *distribution* that is important. See Matthew D. Adler and Eric Posner, "Rethinking Cost-benefit Analysis," *Yale Law Journal* 109 (1999): 165, 170; Amy Sinden, "In Defense of Absolutes: Combating the Politics of Power in Environmental Law," 90 *Iowa Law Review* 90 (2005) 1405, 1413–1416.

32. Cass R. Sunstein, *Risk and Reason: Safety, Law, and the Environment* (Cambridge: Cambridge University Press, 2002), 5–6 ("[the] 'first-generation' debate about whether to base regulatory choices on cost-benefit analysis at all. . . . is now ending, with a substantial victory for the proponents of cost-benefit analysis").

33. See, e.g., Douglas Kysar, *Regulating from Nowhere: Environmental Law and the Search for Objectivity* (New Haven, Conn.: Yale University Press, forthcoming); Shapiro and Schroeder, "Beyond Cost-benefit Analysis"; Ackerman and Heinzerling, *Priceless*; Sidney A. Shapiro and Robert L. Glicksman, *Risk Regulation at Risk: Restoring a Pragmatic Approach* (Stanford, Calif.: Stanford University Press, 2003).

34. See, generally, David M. Driesen, "Is Cost-benefit Analysis Neutral?" *Colorado Law Review* 77 (2006): 335.

35. Shapiro and Schroeder, "Beyond Cost-benefit Analysis," 476.

36. See Thomas O. McGarity, "Presidential Control of Regulatory Agency Decision Making," *American University Law Review*, 36 (1987): 443, 462–463. For the view that tighter presidential control of regulatory decisions is a good thing, see, generally, Kagan, "Presidential Administration."

37. See Shapiro and Glicksman, "Risk Regulation at Risk," 99–100; Ackerman and Heinzerling, *Priceless*, 77–78.

38. Shapiro and Schroeder, "Beyond Cost-benefit Analysis," 455.

39. See ibid.

40. For a fuller explanation of valuing environmental protection, see Verchick, "Feathers or Gold? A Civic Economics for Environmental Law," *Harvard Environmental Law Review* 95 (2001): 101–106.

41. Office of Management and Budget, *Circular A-4, Regulatory Analysis* (2003), www.whitehouse.gov (accessed June 9, 2009), 34.

42. John Applegate et al., *Reinvigorating Protection of Health, Safety, and the Environment* (January 2009), Center for Progressive Reform, www.progressivereform.org (accessed June 9, 2009), 6 (using this example).

43. See W. Kip Viscusi, Rational Discounting for Regulatory Analysis, *University of Chicago Law Review* 74 (2007): 209, 216–217. Louis Kaplow, "Discounting Dollars, Discounting Lives: Intergenerational Distributive Justice and Efficiency," *University of Chicago Law Review* 74 (2007): 79, 84.

44. Martin L. Weitzman, review of *The Stern Review on the Economics of Climate Change, Journal of Economic Literature* 45(Sept. 2007), 703, 713, 716–717.

45. Ackerman and Heinzerling, *Priceless*, 185–86; see also Douglas A. Kysar, "Discounting . . . on Stilts," *University of Chicago Law Review* 74 (2007): 119, 122 (arguing that the defense of long-term discount relies on an unacknowledged, "subtle conceptual shift from individual preferences to collective welfare impacts").

46. Kysar, "Discounting," 124–125.

47. Cass Sunstein, "The Arithmetic of Arsenic," *Georgetown Law Journal* 90 (2002): 2255.

48. See, generally, Robert R. M. Verchick, "The Case against Cost Benefit Analysis," *Ecology Law Quarterly* 32 (2005): 101. Douglas Kysar has called the resources needed for cost-benefit analysis "Herculean." Douglas A. Kysar, "It Might Have Been: Risk Precaution and Opportunity Cost," *Journal of Land Use and Environmental Law* 22, no. 1 (2006): 16. Law professor Richard Parker has discussed the scope of added resources needed to improve cost-benefit analysis. Richard W. Parker, "Grading the Government," *Chicago Law Review* 70 (2003): 1345, 1418 (allowing that enabling the federal government "to perform more—and more thorough—cost-benefit analyses" might require "doubling or tripling" the regulatory budgets of federal agencies).

49. See Stephen Breyer, *Breaking the Vicious Circle: Toward Effective Risk Regulation* (Cambridge, Mass.: Harvard University Press, 1993), 4–5.

50. John D. Graham, "Making Sense of Risk: An Agenda for Congress," in *Risks, Costs, and Lives Saved,* ed. Robert W. Hahn (New York: Oxford University Press, 1996), 183–207.

51. Cass Sunstein, *Worst-case Scenarios* (Cambridge, Mass.: Harvard University Press, 2007), 135 (emphasis added) (citing Elke Weber et al., "Predicting Risk-sensitivity in Humans and Lower Animals: Risk as Variance or Coefficient of Variation," *Psychological Review* 111 [2004]: 430). Sunstein's own views seemed to have evolved on this issue. Compare Cass Sunstein, "Arithmetic of Arsenic," 2255, 2298. (denouncing public fears of arsenic in drinking water as "intuitive toxicology" and implying that overblown fears are of general concern).

52. Posner, *Catastrophe,* 27, table 1.1 (citing *Report of the [U.K.] Task Force on Potentially Hazardous Near Earth Objects,* Sept. 2000, 6) (noting average interval between impacts of a 75-meter "near-earth object" to be 1,000 years); U.S.

Geographical Survey, *Earthquake Probability Mapping* (undated), eqint.cr.usgs .gov (accessed Dec. 29, 2008) (calculating the probability of a magnitude 7 or greater earthquake within 50 kilometers of Manhattan, over a 1,000-year period, to be between 0.025 and 0.030).

53. Clarke, *Worst Cases*, 75.

54. Ibid., 30 (quoting Martin Rees, *Our Final Hour: A Scientist's Warning: How Terror, and Environmental Disaster Threaten Humankind's Future in This Century—On Earth and Beyond* [New York: Basic Books, 2003], 120–121).

55. Margaret Atwood, *Oryx and Crake* (New York: Anchor, 2004).

56. Sunstein, *Worst-case Scenarios*, 138. The text says "1,000 times worse," but this appears to be a typographical error. If my math is right, he means "100,000 times worse."

57. See Ackerman and Heinzerling, *Priceless*, 123–152.

58. Sunstein, *Worst-case Scenarios*, 140–141 (automotive service); A. J. Jacobs, "Fun Couple of the Twenty-first Century," *Esquire*, Oct. 10, 2008 www.esquire.com (accessed Jan. 5, 2009) (quoting Cass Sunstein) (wedding ring).

59. See Posner, *Catastrophe*, 150 (crediting risk aversion in the face of extreme hazard).

60. Weitzman, review of *Stern Review*, 703–724, 715–716.

61. John Rawls, *A Theory of Justice* (1971) (Oxford: Oxford University Press, 1973), 303.

62. Stuart Kauffman, *At Home in the Universe: The Search for the Laws of Self-organization and Complexity* (New York: Oxford University Press, 1995), 302.

63. John Steinbeck, *The Grapes of Wrath* (New York: Viking Press, 1939).

64. See Robert R. M. Verchick, "Dust Bowl Blues: Saving and Sharing the Ogallala Aquifer," *Journal of Environmental Law and Litigation* 14 (1999) 13, 16–17.

65. Woody Guthrie, "Dust Bowl Pneumonia Blues," *Dust Bowl Ballads* (1940), Rounder Records 1988.

66. Steinbeck favored the *Symphony of Psalms*. See Robert DeMott, editor's preface, to John Steinbeck, *Working Days: The Journals of "The Grapes of Wrath"* (New York: Viking, 1989), 13. At other times, Steinbeck found inspiration in Tchaikovsky's *Swan Lake*; see ibid. For more on the work of John Steinbeck and its relevance to environmental policy, see Robert R. M. Verchick, "Steinbeck's Holism: Science, Literature, and Environmental Law," *Stanford Environmental Law Journal* 22 (2003): 3.

67. Shapiro and Schroeder, "Beyond Cost-benefit Analysis," 454 n.118.

68. David E. Adelman, "Scientific Activism and Restraint: The Interplay of Statistics, Judgment, and Procedure in Environmental Law," *Notre Dame Law Review* 79 ( ): 497, 577 (noting that in "Delphi" analysis, scientists' subjectively chosen distributions are "bound to influence greatly, if not determine, the outcome of . . . assessments in fields like environmental science where data are often very limited");

Kysar, "It Might Have Been," 20 (arguing that Monte Carlo techniques depend on certain assumptions about the behavior of unknown probabilities and that when applied to systems following the laws of complexity, Monte Carlo analysis "can lead to dramatically erroneous policy advice, despite the great technological sophistication of the . . . procedure").

69. See, e.g., Aldo Leopold, *The Sand County Almanac, and Sketches Here and There* (1948) (New York: Oxford University Press, 1987).

70. Posner, *Catastrophe*, 141.

71. Sunstein, *Worst-case Scenarios*, 143.

72. See Lawrence Tribe, "Trial by Mathematics: Precision and Ritual in the Legal Process," *Harvard Law Review* 84 (1971): 1329, 1362 ("Readily quantifiable factors are easier to process—and hence more likely to be recognized and reflected in the outcome—than are factors that resist ready quantification").

73. Robert W. Hahn and Cass R. Sunstein, "A New Executive Order for Improving Deeper and Wider Cost-benefit Analysis," *University of Pennsylvania Law Review* 150 (2002): 1489, 1527–1528, 1544.

74. Posner, *Catastrophe*, 176–177.

75. Ibid., 184.

76. Ibid., 174.

77. Lisa Heinzerling, "Accidental Environmentalist: Judge Posner on Catastrophic Thinking," *Georgetown Law Journal* 94 (2006): 853.

78. Sunstein, *Worst-case Scenarios*, 167–168.

79. Daniel A. Farber, "Probabilities Behaving Badly: Complexity Theory and Environmental Uncertainty," *University of California Davis Law Review*, 37 (2003): 145, 155; see also Kysar, "It Might Have Been," 15–16.

9. Mapping Katrina

1. I am indebted to Doug Kysar for sharing with me his thoughts and research on the issues discussed in this section.

2. Professor of Law Douglas Kysar, tenure lecture at Cornell University, Ithaca, New York (2005) (lecture notes on file with the author).

3. Ibid. (citing National Oceanic and Atmospheric Administration, *Revised Standard Project Hurricane Criteria for the Atlantic and Gulf Coasts of the United States* (June 1972) Memorandum HJR-7120.

4. U.S. Army Corps of Engineers, *Performance Evaluation of the New Orleans and Southeast Louisiana Hurricane Protection System: Draft Final Report of the Interagency Performance Evaluation Task Force* (June 2006), ipet.wes.army.mil (accessed Jan. 6, 2009). 1–5, III-12–13.

5. John McQuaid and Mark Schleifstein, *Path of Destruction: The Devastation of New Orleans and the Coming Age of Superstorms* (New York: Little, Brown, 2006), 68.

6. See U.S. Army Corps of Engineers, *Economic and Environmental Principles and*

Guidelines for Water and Related Land Resources Implementation Studies (Mar. 10, 1983), www.usace.army.mil, 32 (describing three "benefit categories" of flood hazard reduction, all based on increasing net income, rather than preserving existing value).

7. See Nancy S. Philippi, *Floodplain Management: Ecological and Economic Perspectives* (Austin, Tex.: R. G. Landes, 1996), 56, 151–52.

8. Bob Marshall, "Levees Get Less Scrutiny than Dams, Critics Note Disparity in Corps' Standards," *New Orleans Times-Picayune*, Mar. 21, 2006, A1 (citing interviews with representatives from the American Society of Civil Engineers and the National Science Foundation).

9. 42 U.S.C. sec. 4331 et seq. (1970).

10. Robert G. Dreher, *NEPA under Siege: The Political Assault on the National Environmental Policy Act*, Georgetown Environmental Law Policy Institute (2005), 65.110.78.8/OurIssues/NEPA/index.cfm (accessed Jan. 11, 2009), 4–7 (providing examples); Council on Environmental Quality, Executive Office of the President, *The National Environmental Policy Act: Study of Its Effectiveness after Twenty-five Years* (Jan. 1997), ceq.eh.doe.gov (accessed Jan. 11, 2009), 8, 18. The Act has proved so attractive that similar laws now exist "in the statute books of 19 states and over 130 nations throughout the world." James Rasband et al., *Natural Resources Law and Policy* (New York: Foundation Press, 2004), 253.

11. Thomas O. McGarity and Douglas A. Kysar, "Did NEPA Drown New Orleans? The Levees, the Blame Game, and the Hazards of Hindsight," *Duke Law Journal* 56 (2006): 179, 187.

12. Ibid.

13. Ibid.

14. Ralph Vartabedian and Peter Pae, "A Barrier That Could Have Been," *Los Angeles Times*, Sept. 9, 2005, A1.

15. Ibid.

16. Ibid.

17. R. Emmett Tyrell, Jr., "Eco-catastrophe Echoes," *Washington Times*, Sept. 16, 2005, A19.

18. Michael Tremoglie, "New Orleans: A Green Genocide," *FrontPageMagazine*, Sept. 8, 2005, frontpagemag.com (accessed Nov. 3, 2008).

19. Dan Eggen, "Senate Panel Investigating Challenges to Levees," *Washington Post*, Sept. 17, 2005, A10.

20. See Ralph Vartabedian and Richard B. Schmitt, "Mid-60s Project Fuels Environmental Fight," *Los Angeles Times*, Sept. 17, 2005, A17.

21. McGarity and Kysar, "Did NEPA Drown New Orleans," 201.

22. Save Our Wetlands, Inc. v. Rush, no. 75-3710, slip op. (E.D. La. 1977), www.save ourwetlands.org (accessed Nov. 3, 2008).

23. Ibid. (emphasis added).

24. McGarity and Kysar, "Did NEPA Drown New Orleans?" 189.

25. U.S. General Accounting Office, *Improved Planning Needed by the Corps of Engineers to Resolve Environmental, Technical and Financial Issues on the Lake Pontchartrain Hurricane Protection Project*, GAO/MASAD-82-39, (Aug. 17, 1982), (archive.gao.gov) 4.

26. Ibid., 7 (reporting a 1982 estimate of $924 million, which factored in estimated costs of inflation through 2008).

27. Ibid. ("Initially the high-level plan was a major competing alternative to the barrier plan; however, it was discarded by the Corps as being too costly").

28. Ibid., 2 (reporting 1981 cost estimates, excluding inflation, for the high-level option and the barrier option of $629 million and $757 million, respectively).

29. U.S. General Accounting Office, *Improved Planning Needed*, 2; ibid.

30. Testimony of Anu Mittal, U.S. Government Accountability Office, before the Subcommittee on Energy and Water Development, Committee on Appropriations, House of Representatives, Sept. 28, 2005, 1 www.gao.gov (accessed Nov. 3, 2008), 1.

31. Ibid.

32. Ibid., 2. The Corps also believed the high-level option would have required more "maintenance" over the long term. ibid. Still, that concern appears to be one of cost rather than safety. In any event, the high-level levees appear to have failed mainly because of poor design and construction, not for lack of maintenance.

33. McGarity and Kysar, "Did NEPA Drown New Orleans," 203.

34. Deborah Mabile Settoon, Citizens for a Safer Jefferson Parish, letter to the editor, *New Orleans Times-Picayune*, June 16, 2007, A5.

35. Save Our Wetlands.

36. Raymond J. Burby, "Hurricane Katrina and the Paradoxes of Government Disaster Policy: Bringing about Wise Governmental Decisions for Hazardous Areas," *Annals of the American Academy of Political and Social Science* 604, no. 1, (March 2006): 171–191.

37. Raymond J. Burby et al., *Flood Plain Land Use Managed: A National Assessment* (Boulder, Colo.: Westview Press, 1985).

10. Planning Our Futures

1. Matthew D. Adler, "Policy Analysis for Natural Hazards: Some Cautionary Lessons from Environmental Policy Analysis," *Duke Law Journal* 56 (2006): 1, 24.

2. Lisa Heinzerling, "Accidental Environmentalist: Judge Posner on Catastrophic Thinking," 94 *Georgetown Law Journal* 94 (2006): 833, 853 (footnotes omitted).

3. Frank Ackerman and Lisa Heinzerling, *Priceless: On Knowing the Price of Everything and the Value of Nothing* (New York: New Press, 2004), 225–226.

4. David A. Dana, "A Behavioral Economic Defense of the Precautionary Principle," *Northwestern Law Review* 97 (2003): 1315, 1331–1333.

5. Daniel A. Farber, "Probabilities Behaving Badly: Complexity Theory and Environmental Uncertainty," *University of California Davis Law Review*, 37 (2003): 145, 167.

6. Ibid., 168.

7. Gill Ringland, *Scenario Planning: Managing for the Future*, 2nd ed. (Chichester, England: Wiley, 2006), 13–15.

8. Ibid., 20–21, 30, 212–213.

9. Ibid., 118.

10. Ibid., 117–123; see also Pieter le Roux et al., "The Mont Fleur Scenarios: What Will South Africa Be Like in the Year 2002?" *Deeper News* 7 (undated) (originally published in *Weekly Mail* [Johannesburg], July 1992), www.gbn.org (accessed May 24, 2009), 7.

11. See "FINSKEN: Scenario Gateway," www.finessi.info (accessed on May 24, 2009); for an overview of the FINSKEN project, see Timothy R. Carter, Stefan Fronzek, and Ilona Bärlund, "FINSKEN: A Framework for Developing Consistent Global Change Scenarios for Finland in the 21st Century," *Boreal Environment Research* 9 (2004): 91–107.

12. Ringland, *Scenario Planning* 117, 121–129.

13. George Sylvester Viereck, *Glimpses of the Great* (London. Macauley, 1930), 377 (citing George Sylvester Viereck, "What Life Means to Einstein," *Saturday Evening Post*, Oct. 26, 1929, 117).

14. T. J. Chermack, "Improving Decision-making with Scenario Planning," *Futures*, 36 (2004): 295–309, sec. 5.1 (noting that planning narratives allow for more complexity and are more memorable).

15. Ibid., sec. 5.2.

16. Ibid., sec. 3.3.2.

17. Jason B. Moats et al., "Using Scenarios to Develop Crisis Managers: Applications of Scenario Planning and Scenario-based Training," *Advances in Developing Human Resources* 10 (2008): 397, 408.

18. There is an important exception. According to the 9/11 Commission report in 1999, a report by the Federal Aviation Administration described the possibility of a "suicide hijacking operation" involving Osama Bin Laden. But the agency's analysts judged the scenario unlikely, and the CIA did not follow up with an analytic assessment. National Commission on Terrorist Attacks upon the United States, *9/11 Commission Report* (2004), www.9-11commission.gov (accessed May 24, 2009), 345.

19. Ibid., 297.

20. Moats et al., "Using Scenarios," 410–411.

21. Ibid., 413.

22. Ibid., 417; Chermack, "Improving," sec. 6. For a practical guide on scenario planning for use in public policy, see Gill Ringland, *Scenarios in Public Policy* (Chichester: Wiley, 2002).

23. 40 C.F.R. sec. 1502.22(b) (2003) (emphasis added); see also Cass Sunstein, *Worst-case Scenarios* (Cambridge, Mass.: Harvard University Press, 2007), 19–21 (discussing this example); Farber, "Probabilities Behaving Badly," 166–167 (same).

24. Ibid.

25. Daniel A. Farber, *Confronting Uncertainty under NEPA*, May 13, 2009, unpublished manuscript, available on Social Science Research Network, Identification no. 1403723, ssrn.com (accessed May 26, 2009), 28.

26. Millennium Ecosystem Assessment, *Ecosystems and Human Well-Being: Synthesis* (Washington D.C.: Island Press, 2005).

27. Ibid.

28. See U.S. Army Corps of Engineers, *Louisiana Coastal Protection and Restoration: Draft Technical Report* (Feb. 2008), lacpr.usace.army.mil (accessed Dec. 28, 2008), 3.

29. Ibid.

30. U.S. Army Corps of Engineers, *Louisiana Coastal Protection and Restoration Technical Report (Draft)*, (Feb. 2008), lacpr.usace.army.mil (accessed Jan. 14, 2008).

31. Ibid., 121, table 7–2.

32. See ibid., 43–45.

33. Andrew Bridges, "Development Raises Flood Risk across U.S.," Associated Press, Feb. 18, 2006, news.yahoo.com (accessed Nov. 3, 2008). In a nationwide levee review, mandated by Congress after Katrina, the U.S. Army Corps of Engineers recently released a list of 122 levees in 30 states and territories that it says have an unacceptable risk of failing. U.S. Army Corps of Engineers, *Levees of Maintenance Concern* (Feb. 1, 2007), www.hq.usace.army.mil (accessed Nov. 3, 2008). Notably, 38 are located in California—nearly all in the state's "Sacramento District"; other jurisdictions with multiple listings are Alaska, Arkansas, Connecticut, the District of Columbia, Hawai'i, Idaho, Illinois, Kentucky, Maryland, Massachusetts, Michigan, New Mexico, Oregon, Pennsylvania, Puerto Rico, Rhode Island, and Washington; ibid.

Conclusion

1. For a favorable review of this production, see David Cuthbert, "'Godot' Is Great," *New Orleans Times-Picayune*, Nov. 6, 2007 (with photographs of the performance), www.blog.nola.com (accessed Jan. 15, 2009).

2. Nicholas Stern, *The Economics of Climate Change: The Stern Review* (Cambridge: Cambridge University Press, 2007), www.hmtreasury.gov.uk (accessed Jan. 12, 2009), 56.

3. See Matthew E. Kahn, "The Death Toll from Natural Disasters: The Role of Income, Geography, and Institutions," *Review of Economics and Statistics* 87, no. 2 (2005): 271–284, 271.

4. Stern, *Stern Review*, 32, 52.

5. Ibid., 32, 52.

6. Ibid., 110, 114.

7. Ibid., 114.

8. Ibid.

9. Ibid.

10. See generally, Committee on Environment and Natural Resources National Science and Technology Council, *Scientific Assessment of the Effects of Global Change on the United States* (undated), www.climatescience.gov (accessed June 3, 2008).

11. Ibid., 14, 165, 167–168.

12. Samuel Beckett, *Waiting for Godot* (New York: Grove Press, 1954), 51.

INDEX

Abramowitz, Janet, 26, 27, 33, 47
Accountability, of agencies, 176
Accretion, island building by, 35
Ackerman, Frank, 220, 237–238
Action, disaster policy and, 251–254
Activists, environmental, 123–124
Adaptive system, in Louisiana, 74
Administrative Procedures Act, 66
Adverse possession doctrine, 88
Africa, coastal islands in, 35
African Americans: exposure to disasters, 131–135;
    after Florida hurricane of 1928, 109–110; in
    Galveston, 143; after Mississippi River flood
    (1927), 108–109; poverty and, 136–139;
    psychosocial trauma to elderly, 136; racism
    in Katrina aftermath and, 137; transportation
    for, 121; vulnerability to disasters, 135–136.
    See also Bias
African Commission on Human and Peoples'
    Rights, 189
Agencies. See Government agencies
Agency for International Development (U.S.),
    189
Air pollution, suspension of standards, 154. See
    also Climate change; Pollution
Alabama, Emelle, 124; OCS revenues in, 85;
    public trust doctrine in, 96
Alaska, coastline and lowland riverbanks of, 30, 93
Allocation of resources, 57; neoliberalism and,
    150
American Creosote Works, 169–170
"America's Wetland" campaign (Louisiana), 86
Anthropogenic climate change, 6
Antidiscrimination law, 170–173
Antiurban policies, 150–151

Army Corps of Engineers: Comprehensive
    Master Plan for Sustainable Coast, 71;
    cost-benefit analysis by, 202, 225–226;
    Hurricane Betsy and, 224; Hurricane Katrina
    and, 2, 3; levee systems and, 21, 229; Louisiana
    coastal wetlands and, 19, 20; post-Katrina storm
    protection and, 246–247; Save Our Wetlands
    litigation and, 231; water management by,
    67–70; wetland maintenance and, 79, 87. See
    also Barrier option; Levees
Arnold, Joseph, 69
Arrow, Kenneth, 238
Arsenic, removal of, 134–135
Arts, catastrophes and, 105–106
Arugam Bay, Sri Lanka, 152–154, 155, 188
Asian Americans, in Gulf region, 142
Asian tsunami (2004), 5, 26, 38, 112, 253; recovery
    activities after, 42; Sri Lanka development and,
    153, 188; women after, 139
Assets: protection of, 53–54, 90–93; resources as, 49
Atchafalaya River region: coastal wetlands in, 16;
    storm protection in, 70–79
Atlantic seaboard region, race in, 131
Atwood, Margaret, 210
Australia, Great Barrier Reef in, 36
Avenal case, 99, 100

Baker, Richard, 155
Banaji, Mahzarin, 162
Bandelier National Monument (New Mexico),
    179
Bangladesh: mangrove forests in, 32; risk
    reduction in, 184–185
Barasich v. Columbia Gulf Transmission
    Company, 90–93